Lowering of the boiling curve of biodiesel by metathesis

Final report on the project

Lowering of the boiling curve of biodiesel by metathesis

Project manager: Prof. Dr.-Ing. Axel Munack

Participating staff members of the Johann Heinrich von Thünen Institute:
Dipl.-Ing. (FH) Christoph Pabst
Dipl.-Ing. Barbara Fey
Dipl.-Chem. Kevin Schaper
Dipl.-Chem. Lasse Schmidt
Dipl.-Chem. Olaf Schröder

Prof. Dr. Michael A. R. Meier
Rowena Montenegro
Karlsruhe Institute of Technology

Prof. Dr.-Ing. Peter Eilts
Dipl.-Ing. Michael Kaack
Institute of Internal Combustion Engines, Technical University of Braunschweig

Prof. Dr. med. Jürgen Bünger
Steinbeis Transfer Centre for Biofuels and Environmental Measurement Technology, Coburg

and also
Dipl.-Ing. (FH) Alexander Mäder
Prof. Dr. Jürgen Krahl
Technology Transfer Centre Automotive, University of Applied Sciences and Arts, Coburg

Promoted by the Union zur Förderung von Oel- und Proteinpflanzen e.V. (UFOP)
(Union for the Promotion of Oil and Protein Plants)
under project number 540/085

Braunschweig, Coburg and Karlsruhe in June 2012

Institute for Agricultural Technology and Biosystems Engineering

Bundesallee 50
38116 Braunschweig

Tel. +49 (0)531 · 596-4201
Fax +49 (0)531 · 596-4299

www.vti.bund.de
ab@vti.bund.de

Bibliografische Information der Deutschen Nationalbibliothek
Die Deutsche Nationalbibliothek verzeichnet diese Publikation in der
Deutschen Nationalbibliografie; detaillierte bibliografische Daten sind im Internet
über http://dnb.d-nb.de abrufbar.
1. Aufl. - Göttingen : Cuvillier, 2014

978-3-95404-554-9

© CUVILLIER VERLAG, Göttingen 2014
Nonnenstieg 8, 37075 Göttingen
Telefon: 0551-54724-0
Telefax: 0551-54724-21
www.cuvillier.de

Contents

Table of Figures

Index of Tables

Abbreviations

Ace	Acenaphthene
Acet	Acetaldehyde
Acro	Acrolein
AdBlue	32.5% aqueous urea solution
EGR	Exhaust gas recirculation
Ant	Anthracene
B20	Blend of 20% rapeseed oil methyl ester and 80% diesel fuel
BaA	Benz[a]anthracene
BAM	Bundesanstalt für Materialforschung (Federal Institute for Materials Research and Testing)
BaPyr	Benzo[a]pyrene
BbFla	Benzo[b]fluoranthene
Benz	Benzaldehyde
BHT	Butylhydroxytoluene
BkFla	Benzo[k]fluoranthene
Buton	2-Butanone
Buty	Butyraldehyde
BS	British Standards
C_3H_8	Propane
Cat.	Catalyst
CFPP	Cold Filter Plugging Point (temperature threshold value of filterability)
Chr	Chrysene
CI-MS	Mass spectrometer with chemical ionisation
CEC	CEC legislative fuel RF-06-03
CLD	Chemiluminescence detector
CM_1 to CM_6	Designation of cross-metathesis products
CO	Carbon monoxide
$CO(NH_2)_2$	Urea
CO_2	Carbon dioxide
Crot	Crotonaldehyde
DAD	Diode array detector
DBAnt	Dibenzo[a,h]anthracene
DF	Diesel fuel
DEE	Diesel engine emissions
DMSO	Dimethylsulphoxide
DNA	Deoxyribonucleic acid
DNPH	2,4-Dinitrophenylhydrazine
EC	European Community
ELPI	Electrical low-pressure impactor
EPA	United States Environmental Protection Agency
eq	Molar equivalent
ESC	European Stationary Cycle

ETC	European Transient Cycle
FAME	Fatty acid methyl ester
FID	Flame ionisation detector
Fla	Fluoranthene
Flu	Fluorene
Form	Formaldehyde
GC	Gas chromatography
GC-MS	Gas chromatography with mass spectrometer
H_2O	Water
H_2SO_4	Sulphuric acid
HC	Hydrocarbons
HDPE	High density polyethylene
Hexa	Hexaldehyde
HFRR	High-frequency reciprocating rig (value for lubricity)
HNCO	Isocyanic acid
HNO_3	Nitric acid
HPLC	High-performance liquid chromatography
HVO	Hydrotreated vegetable oil
ICP-MS	Mass spectrometry with inductively-coupled plasma
IPyr	Indeno[1,2,3-cd]pyrene
M20	Metathesis blend of 20% metathesis fuel and 80% diesel fuel
MA to MN	Designation of metathesis products in sequence of manufacture
Meth	Methacrolein
MMS	Methyl methanesulphonate
Na_2SO_4	Sodium sulphate
Nap	Naphthalene
NH_3	Ammonia
NO	Nitrogen monoxide
NO_2	Nitrogen dioxide
NO_x	Nitrogen oxides ($NO+NO_2$)
O_3	Ozone
OEM	Original Equipment Manufacturer
PA	Polyamide
PAK	Polycyclic aromatic hydrocarbons
PE	Polyethylene
PGV	Particle size distribution
Phe	Phenanthrene
PM	Particle mass
Prop	Propinonaldehyde
PTFE	Polytetrafluoroethylene
Pyr	Pyrene
RME	Rapeseed oil methyl ester
RT	Room temperature

SCR	Selective catalytic reduction
SCR-Cat.	SCR catalyst
SimDis	Simulated distillation
SM_1 to SM_6	Designation of self-metathesis products
SMPS	Scanning Mobility Particle Sizer
THF	Tetrahydrofuran
Tolu	m-Tolualdehyde
TON	Turnover number
Valer	Valeraldehyde
2-AF	2-Aminofluorene
3-NBA	3-Nitrobenzanthrone

Summary

For this project, rapeseed oil methyl ester (RME) was adapted to match the properties of diesel fuel (DF) by a metathesis reaction. In particular, this adaptation concerned the boiling property. The resultant fuel was very similar to the diesel fuel with respect to its boiling curve. However, the metathesis fuels still contained the ester groups that are already present in biodiesel. In the course of the project, twelve different metathesis products were produced and examined with respect to their properties and the exhaust emissions resulting from engine combustion.

As a design instrument for the adjustment of the boiling curve of a fuel, the alteration of biodiesel molecules by metathesis proved to be extremely effective. Beginning with rapeseed oil methyl ester, which predominantly comprises oleic, linoleic and linolenic acids, a variation in the chain length, which substantially influenced the boiling curve of the fuel produced, was achieved through the implementation of catalysts with the help of the metathesis reaction and the use of 1-hexene. In this way, it was possible to approximate the boiling behaviour of the RME to that of the DF over a wide range. Very different products could be created through variation of the catalysts and the ratio of biodiesel to 1-hexene. As a 20% blend with diesel fuel some of these could hardly be differentiated from DK, based on the boiling curve.

Tests were carried out on the newly-produced fuel samples to determine their miscibility with other fuels and with engine oil. No problems of any kind were encountered when mixing with further fuel components such as DF and HVO (hydrotreated vegetable oil). However, as already described by Krahl et al., 2011, turbidity resulted upon addition of aged RME that could not be ascribed to the metathesis products. Also with respect to behaviour in engine oil, no problems of any kind, such as the formation of precipitates, occurred at temperatures between -18 °C and 90 °C during a storage period of 24 hours.

In addition to behaviour in comparison with other operating materials, first material compatibility tests were carried out on two selected polymers. The biodiesel resistant materials PA 66 (Polyamide) and HDPE Lupolen 4261 (High Density Polyethylene) were also resistant to the metathesis fuel (MA). The expansion behaviour was in the order of magnitude of RME. The mass increase of HDPE in metathesis fuel and RME was in fact clearly lower than in DF. For PA, the mass of the samples increased by almost 10% on storage in fuel for seven days at 70 °C. With the removal of HDPE, the mass increase was actually much lower at less than 0.1%. In the tensile test, no significant differences resulted between the samples that were stored in the different fuels. The modulus of elasticity of the PA samples was about 2000 MPa and about 500 MPa for HDPE. The tensile strength was also of the same order of magnitude for all samples at about 60 MPa for PA and 25 MPa for HDPE. Hence the first tests did not lead to an expectation of disadvantages in the material compatibility of plastics and metathesis fuel. Even the corrosive effect on copper was within the range of the current diesel fuel norm (Table 4-4).

In addition to the determination of these fundamental fuel characteristics, exhaust gas analyses also took place in engine tests with blends of 80% fossil diesel fuel and 20% metathesis components. The first measurements were carried out on the single-cylinder test engine. In the tests, significant differences between the different metathesis products were seldom encountered. Two of the compared fuels were selected for further work using a matrix that took both the limited emissions and also the boiling curve and the biodiesel proportion into account. More comprehensive testing took place using an OM 904 LA commercial vehicle engine and evaluation of the emissions and combustion behaviour using an AVL single-cylinder research engine based on a MAN D28.

In addition to the regulated emissions NO_x, PM, HC and CO, the non-regulated exhaust gas components ammonia, polycyclic aromatic hydrocarbons, carbonyls, mutagenicity and the particle size distribution were tested using the OM 904 LA. DK and RME were used as comparison fuels. Due to their blend ratio, the two tested metathesis fuels were directly compared to a B20 blend of the two comparison fuels.

In operation with metathesis fuel blends, the emissions of the OM 904 LA only showed very slight deviations from B20. The nitrogen oxide emissions for RME were much higher than for DF, and also the mixed fuels with RME or metathesis proportion showed a slight increase. The opposite effect was observed with particle mass, where the use of RME led to a reduction of 25%. However, this trend was not indicated by the blends. Their particle masses lay within the order of magnitude of DF at 0.01 g/kWh. A significant reduction in hydrocarbon and carbon monoxide emissions was also observed for RME. For the mixtures, a reduction was only found with the HC emissions. One of the metathesis components displayed slight advantages in terms of carbon monoxide emissions, since no increase in CO emissions was determined here compared to DF, as with the other blends. However, almost all regulated emissions lay within the Euro IV norm that applies to the engine. All specifications were attained for four of the five fuels used. Only the nitrogen oxide emissions of RME slightly exceeded the threshold value of 3.5 g/kWh.

There was a small difference in particle size distribution. Here, the metathesis fuels compared to the B20 blend displayed a slight increase in particle count in the size range from 28 nm to 1000 nm. In the larger range from 1 µm to 10 µm, the values for B20 were then substantially higher. Even in the case of the other tested non-regulated exhaust gas components, there were only small differences between the metathesis fuel blends and B20. The carbonyls were also in the same order of magnitude and no significant differences could be discerned between the fuels used. With regard to mutagenicity, it was clear that the use of a SCR catalytic converter led to such low emissions of mutagenic substances that slight mutagenic tendencies in the emissions could only still be measured with DF.

The tests of the metathesis fuels in the AVL single-cylinder research engine based on the MAN D28 regarding the combustion and emissions behaviour displayed no important differences with respect to the comparison fuels, within the limits of the precision of the available measuring technology. With the regulated emissions, small advantages were discerned for the metathesis fuel blends over DF with respect to PM, HC and CO. The behaviour was the exact opposite with nitrogen oxide emissions.

In summary, within the framework of the implemented tests, no evidence could be found that contradicted the suitability of the metathesis fuels for engine combustion.

1 Introduction

For a long time now, increasing the renewable components in fuel has been of as much interest as the reduction of emissions in motor vehicles. In order to increase the proportion of renewable energies and to reduce dependency on mineral oil, there is great political interest in the addition of biogenic components to fossil fuels. Diesel fuel currently makes up a quota of 7% (BioKraftQuG, 2006). A renewable energy proportion of 10% has been prescribed for the transport sector in the year 2020 by the European Union (EC guideline 2009/28).

In addition, the tightening of emissions standards places increasingly high demands on vehicle manufacturers, and also especially on the quality of the fuels used. The additive component that is currently predominantly used for diesel fuels is biodiesel. However, in its current form it is not optimally suited to modern car engines. A primary reason for this is its boiling behaviour, which for example leads to problems such as oil dilution by the addition of fuel to the engine oil, especially in the case of after-injection for the regeneration of particle filters (Tschöke et al., 2008). The biodiesel contained in engine oil can then form oligomers and polymers that deposit as oil sludge. For this reason, the boiling behaviour of biodiesel was changed in this project so that the fuel can evaporate more easily from the engine oil due to its lower boiling point. In order to attain this adaptation of the boiling curves to match that of the fossil diesel fuel, the properties of rapeseed oil methyl ester were altered both with self-metathesis and also cross-metathesis. In particular, new fuels are to be produced by homogeneous and heterogeneous catalysed cross-metathesis of fatty acid methyl esters with linear olefins that display behaviour in engine operation similar to fossil diesel fuel without allowing oil dilution and oil sludge formation. These possibilities of fuel design could form the basis for a substantial increase in the renewable proportion in diesel fuel without having negative effects on the engine. In this way, the use of plant oils and biodiesel as a fuel could be expanded.

In order to be able to evaluate the behaviour of the altered fuels, analyses of the boiling point and evaluation of the emissions in engine operation are required. Here, the emissions behaviour can not only be described by determination of the amounts of statutorily regulated exhaust gas components nitrogen oxide, particle mass, carbon monoxide and hydrocarbons. Measurement of the statutorily non-regulated components are also necessary for the estimation of the emissions behaviour and any resultant risks, since as is the case with polycyclic aromatic hydrocarbons or mutagenicity in the Ames test, they provide indications for the effect on health of diesel engine emissions. This cannot be derived from the statutorily regulated components. In addition, consideration of the particle concentration and the ammonia emissions is also necessary as these will belong to the regulated components in future (EC Regulation No. 582/2011).

In summary, the project demonstrates a sustainable development approach for fuels since the fuel design has been accompanied by engine and exhaust analysis tests right from the start.

2 Production and basic characterisation of metathesis fuels

The metathesis fuels were produced at the Karlsruhe Institute for Technology. The chemical modification of biodiesel took place using a known catalytic reaction, olefin metathesis, to reduce the boiling curve of biodiesel. This was achieved by cross-metathesis reactions of biodiesel with 1-hexene (and other olefins), which led to a shortening of the alkyl chain of the fatty acid methyl esters. Biodiesel that was synthetically modified in this way produced a boiling curve similar to that of diesel fuel extracted from mineral oil.

2.1 Starting materials and analytical process

a) Starting materials

- Rapeseed oil methyl ester with a fatty acid spectrum comprising 4.7% methyl palmitate, 1.6% methyl stearate, 61.2% methyl oleate, 19.2% methyl linoleate and 10.1% methyl linolenate
- 1-hexene, 97%

b) Catalysts

- Commercially-available ruthenium catalysts (see Fig. 2-1) from Sigma-Aldrich and Umicore were used

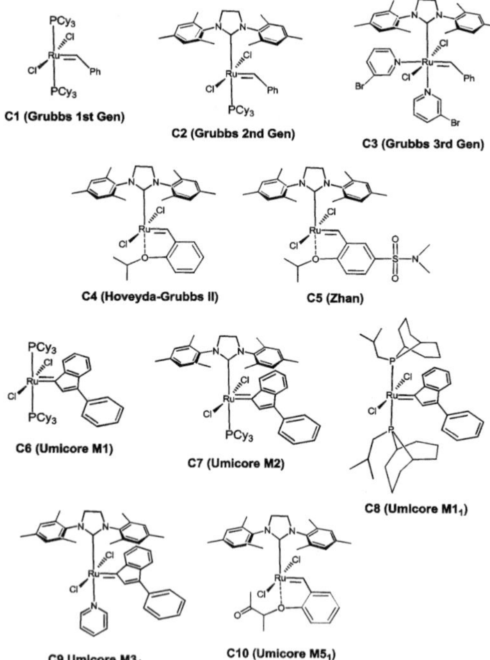

Fig. 2-1: Ruthenium-based olefin metathesis catalysts used for the initial screening

c) Analysis

- Examination of the hydrocarbon composition of the samples of reaction solutions via GC (Bruker 430-GC) and/or GC-MS (Varian-431-GC/MS instrument)
- Simulated distillations (SimDis) of the metathesis samples according to the method ASTM D7169 via GC (ASG, Neusäss and TAC, Coburg)
- Determination of the ruthenium content of the synthetically modified fuels via ICP-MS (Varian 820) using ruthenium standards with 1 to 10 ppm (ASG, Neusäss, Deutschland)

2.2 Metathesis reactions

2.2.1 Cross-metathesis of biodiesel with 1-hexene

Cross-metathesis reactions of biodiesel and 1-hexene were carried out with 5 ml biodiesel, the corresponding quantity of 1-hexene and tetradecane (internal standard) in a Radleys carousel 12 Plus^{TM} carousel reactor with continuous stirring. To simplify calculation of the required quantities of reactants, it was assumed that the biodiesel only comprised methyl oleate. Reactions with catalysts **C1-C5** were carried out at 40 °C, whilst reactions with **C6-C10** took place at the somewhat higher temperature of 50 °C (it is known that C6-C10 have slightly higher initiation temperatures). The reaction temperatures were selected low to prevent 1-hexene (boiling point: 63 °C) from evaporating. After a t=0 sample was taken, in each case one of the catalysts **C1–C10** (0.02 to 0.10 mol% based on the quantity of biodiesel) was added to the reaction mixture. All reactions ran for four hours and were replicated. Samples were taken at regular intervals for GC or GC-MS analyses. In order to stop the reaction before the GC and GC-MS analysis, an excess of vinyl ethylether (50 mol%) was added to the reaction mixture.

The consumption of methyl oleate and methyl linoleate yielded a mixture of self and cross-metathesis products comprising alkenes, dienes, esters and diesters of different chain length (Fig. 2-2). Methyl oleate and linoleate had a proportion of 80.4% of the unsaturated fatty acids in the utilised biodiesel in a ratio of 3:1. The entire fatty acid profile can be found in Table 6-1 in the appendix. The possible resultant compounds are designated as self-metathesis product (SM$_1$ to SM$_6$) and cross-metathesis product (CM$_1$ to CM$_6$). Here, it should be noted that after a metathesis reaction all double bonds are present as cis / trans mixtures, as shown in Fig. 2-2. The compounds produced in this way have clearly different boiling points and should therefore render a modified boiling curve compared to that of the starting product (biodiesel).

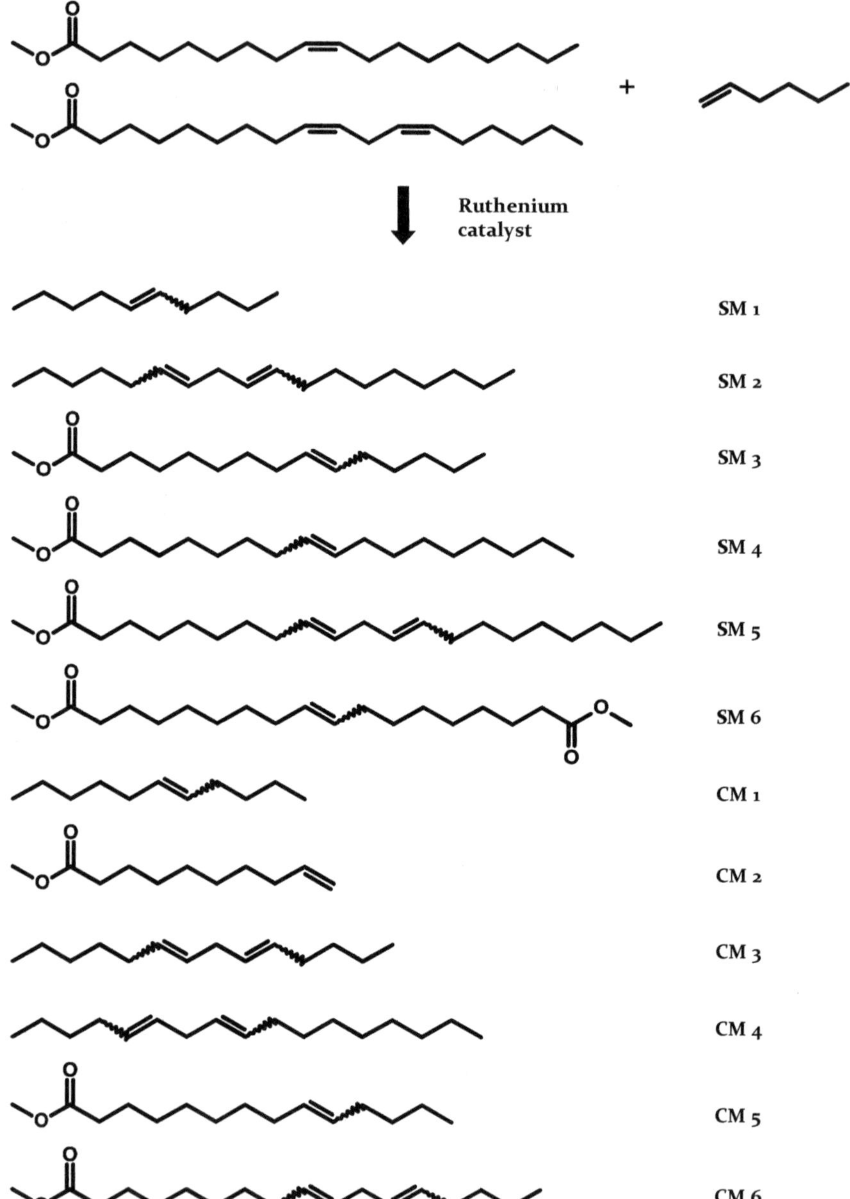

Fig. 2-2:   Product mixture resulting from cross-metathesis of biodiesel with 1-hexene

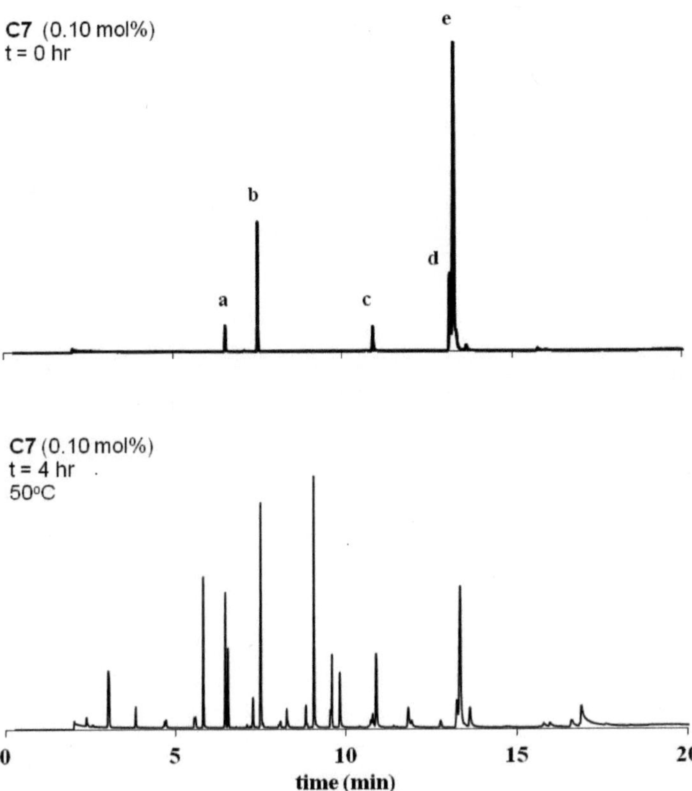

Fig. 2-3: GC-MS of biodiesel before the addition of catalyst (top) and after the cross-metathesis reaction of bio-diesel with 1-hexene (bottom); (a) tetradecane, (b) BHT, (c) methyl palmitate, (d) methyl linoleate, (e) methyl oleate

Fig. 2-3 shows a representative GC-MS chromatogram of biodiesel before and after the addition of one of the catalysts (**C7**). At t=0 only tetradecane (standard), BHT (antioxidation agent in the THF used for sample preparation) and the fatty acid methyl ester of the biodiesels was detected. After four hours' reaction time, the cross-metathesis yielded new products of different chain length, as expected and desired.

In order to identify the most suitable catalyst for this reaction, all catalysts shown in Fig. 2-1 were first screened. The catalyst loading was initially set to 0.10 mol% and tested with a mixture of biodiesel and 1-hexene in a molar ratio of 1:1. All reactions were replicated and displayed good reproducibility.

Fig. 2-4 shows the results of this catalyst screening. As clarified in the diagram, **C2**, **C4**, **C7**, **C9** and **C10** allow almost complete conversion even at the low catalyst loading of 0.10 mol%. These catalysts were then used for the next phase of testing and further optimisation tests.

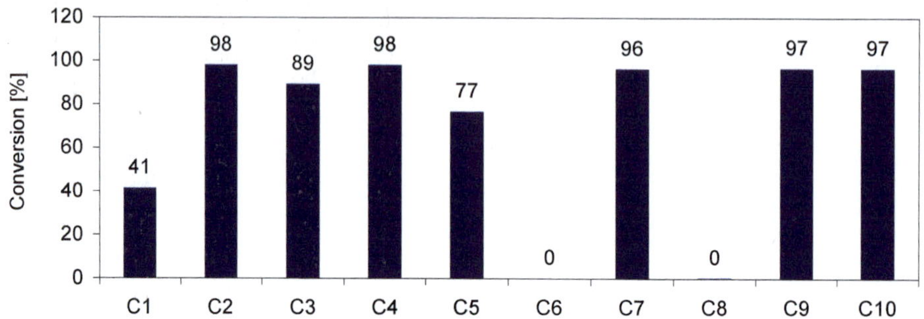

Fig. 2-4:   Results of the catalyst screening (conversion calculated based on all unsaturated FAME) for the cross-metathesis of biodiesel with 1-hexene with 0.10 mol% catalyst loading

The catalyst loading was then reduced to quantities from 0.05 to 0.02 mol% in increments of 0.01 mol%. At 0.05 mol%, **C2**, **C4** and **C10** were still very active; however, at 0.04 mol% and lower loading, these catalysts exhibited incomplete conversions (Fig. 2-5). Due to the comparatively low level of conversion with **C7** and **C9**, and due to the clear reduction in conversion with **C2**, these were not considered for further tests.

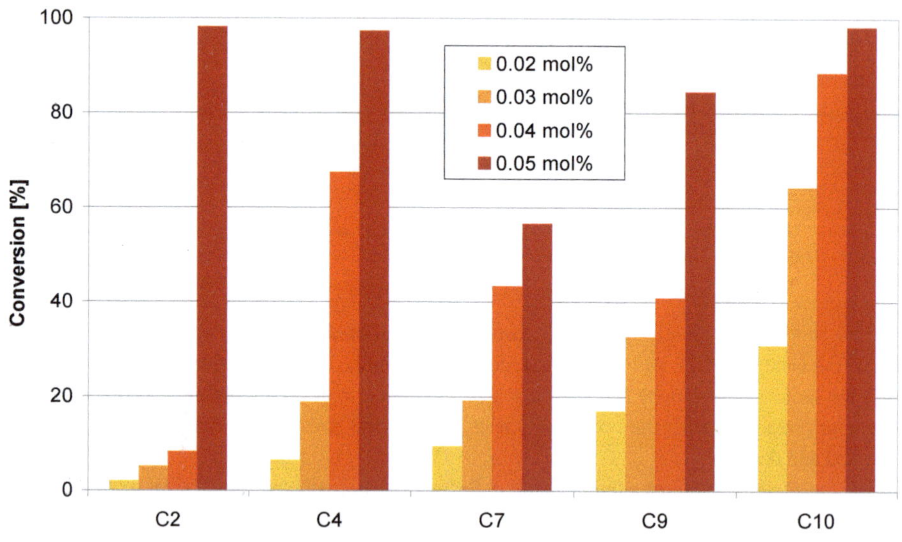

Fig. 2-5:   Cross-metathesis of biodiesel with 1-hexene with low catalyst loading

The two most active catalysts, **C4** and **C10**, also exhibited incomplete conversion with lower catalyst loading, although much less strongly pronounced. In order to develop a better understanding of this process and the activity of the catalysts, the comparison of turnover number (TON) is better suited than just the conversion.

Catalyst loading [mol %]	Catalyst	Conversion [%]	TON[a]
0.04	**C4**	67.3	1682
	C10	88.4	2209
0.03	**C4**	18.7	623
	C10	64.1	2138
0.02	**C4**	6.32	316
	C10	30.7	1534

[a] TON calculated as % conversion / % mol catalyst loading (Ngo et al., 2006)

Table 2-1 shows that **C10** attains TONs of > 2000 for direct conversion of unpurified biodiesel with extremely low catalyst loading.

Catalyst loading [mol %]	Catalyst	Conversion [%]	TON[a]
0.04	**C4**	67.3	1682
	C10	88.4	2209
0.03	**C4**	18.7	623
	C10	64.1	2138
0.02	**C4**	6.32	316
	C10	30.7	1534

[a] TON calculated as % conversion / % mol catalyst loading (Ngo et al., 2006)

Table 2-1:   Comparison of conversion figures (TON: turnover number)

A series of metathesis reactions was then carried out to test the effects of different ratios of 1-hexene to biodiesel. The quantity of biodiesel was kept constant at 100 mol%, whilst the quantity of 1-hexene was varied from 0 to 100 mol% (in increments of 20 mol%). As expected, higher quantities of 1-hexene produced increased quantities of cross-metathesis products.

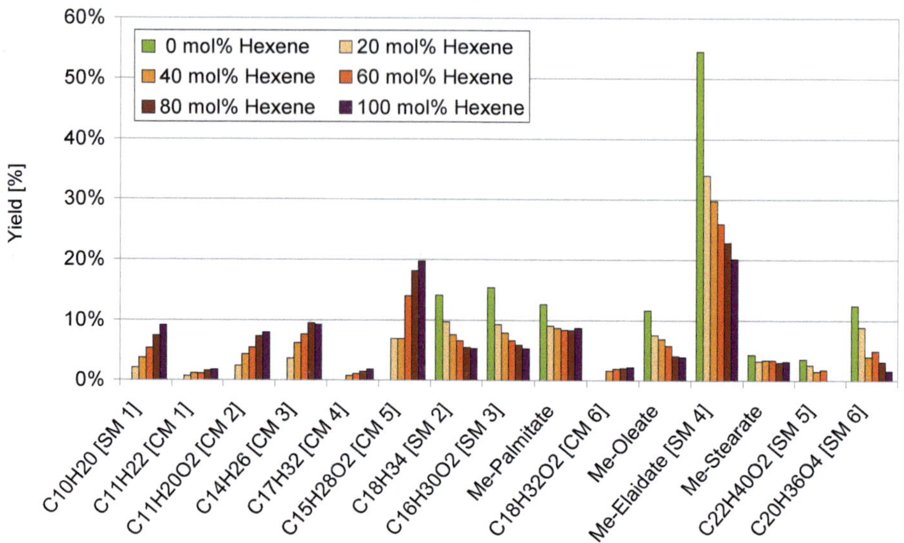

Fig. 2-6:   Composition after the metathesis reaction with different quantities of 1-hexene (data determined using GC)

The opposite was observed for the self-metathesis products of biodiesel. The quantity of unsaturated FAME remained constant, whilst methyl oleate was partially converted. Methyl-linoleate was completely consumed in the reactions. Fig. 2-6 shows very clearly that more of the low-molecular cross-metathesis products were formed with increasing quantities of 1-hexene.

These mixtures (Fig. 2-6) were then tested via simulated distillation (SimDis) to determine their boiling curve (Bachler et al., 2010).

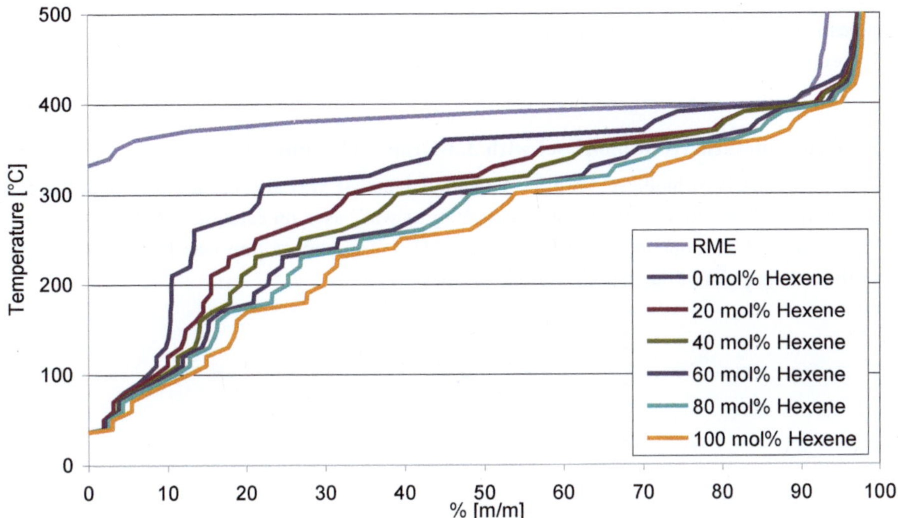

Fig. 2-7:   Results of the simulated distillation (SimDis) of synthetically modified biodiesel (ASG, Neusäss)

As shown in Fig. 2-7, increased quantities of 1-hexene had an increasing positive effect on the boiling curve of the synthetically-modified biodiesel and an increasingly steady boiling curve results. This behaviour can be explained by the generation of compounds with low molecular mass (and hence lower boiling point) as mentioned above during the cross-metathesis and correlates very well with the compositions of the mixtures as presented in Fig. 2-6.

The distillation curve of biodiesel with 100 mol% hexene is an almost continuously increasing curve that is much closer to a conventional diesel fuel than the original biodiesel. This can be described with the generation of a good distribution of light and heavy volatile substances. In comparison with the boiling curve of pure biodiesel, even the self-metathesis curve shows more of a continuous profile. As the curves in between show (20 mol% to 80 mol% 1-hexene), it is possible to determine the fuel properties by a simple variation of the ratio between biodiesel and hexene.

2.2.2 Cross-metathesis of biodiesel with limonene and pinene

The cross-metathesis of biodiesel was also tested with the naturally-occurring compounds limonene and pinene. The reactions were tested with the three most active Umicore catalysts (Umicore M2, Umicore $M3_1$ and Umicore $M5_1$) of a concentration of 0.10 mol% at a temperature of 50 °C.

(L)-(-)-limonene   (S)-(-)-limonene   (-)-β-pinene

However, the GC results showed that the resultant products were largely those of the self-metathesis of biodiesel. Since no cross-metathesis reactions of biodiesel with limonene or pinene were observed, these reactions were not further tested with other catalysts.

2.2.3 Cross-metathesis of biodiesel with 3,3-Dimethyl-1-butene

The cross-metathesis of biodiesel with 3,3-Dimethyl-1-butene was also carried out using the most active catalysts. Evaluation of the most active catalysts was based on catalyst screening (Grubbs II, HG II, Umicore M2, Umicore $M3_1$ and Umicore $M5_1$). The temperature was fixed at 40 °C, just under the boiling point of the dimethylbutene (41 °C).

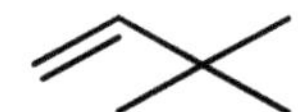

3,3-Dimethyl-1-butene

Although high levels of conversion (> 96%) were observed for all catalysts (except Grubbs II, 42% conversion) with just 0.10 mol% catalyst loading, the GC analysis showed that only self-metathesis products of biodiesel were produced.

2.2.4 Cross-metathesis of rapeseed oil with 1-hexene

Cross-metathesis tests were also carried out on 1-hexene (in excess) with rapeseed oil that was bought in a supermarket. The rapeseed oil was used without further cleaning and since no trans-esterification took place before use, the cross-metathesis reactions with Hoveyda-Grubbs II and Umicore $M5_1$ (0.10 mol%) at temperatures of 40 °C and 50 °C in the GC analysis only exhibited products with low molecular weight in the range from C_{10} to C_{14}.

2.3 Optimisations

2.3.1 Temperature variation and method of catalyst addition

In addition to optimisation of the reaction conditions described above, tests were carried out to establish whether a further reduction of the required quantity of catalyst is possible by increasing the temperature to 80 °C and 100 °C and/or by the method of catalyst addition.

Delayed addition (by hand):	The total amount of catalyst was divided into four equal parts and added manually at intervals of one hour respectively.
Delayed addition (pump):	A pump delivers the catalyst to the reaction mixture at constant dosage speed; the catalyst is diluted with toluene.
Immediate addition:	Complete addition of the catalyst at the start of the reaction.

Catalyst	Temp [°C]	Method	Conversion [%]	TON
Umicore M5$_1$	80	immediate	6.79	679
		manual delayed	9.89	989
		pump delayed	31.65	3165
	100	immediate	3.31	331
		manual delayed	30.60	3060
		pump delayed	50.75	5075
Hoveyda-Grubbs II	80	immediate	0.38	380
		manual delayed	57.20	5720
		pump delayed	84.10	8410
	100	immediate	2.97	297
		manual delayed	81.62	8162
		pump delayed	93.12	9312

Table 2-2: Comparison of conversions and TONs

Table 2-2 compares the results of the various addition methods using the example of the self-metathesis of biodiesel with the two most active catalysts. The data clearly show that slow addition drastically increases the conversion in the metathesis reaction, and that this effect is strongest when dosed continuously with a pump, whereby very high TONs of 9000 are attained. Similar effects are known from literature for other metathesis reactions (Miao et al., 2011; Skaanderup and Jensen, 2008).

2.3.2 Biodiesel pre-treatments

It was investigated as to whether pre-treatment of biodiesel would improve the catalyst results, since any impurities in the biodiesel could deactivate the catalyst. The following substances were used for the pre-treatment:

Silica gel, aluminium oxide (ALOX), activated charcoal and Magnesol

Silica gel / ALOX: Biodiesel was filtered over silica gel or ALOX

Activated charcoal: Biodiesel was stirred with activated charcoal for different time periods (1h, 4h, overnight). The activated charcoal was then removed by filtration

Magnesol: Magnesol (1 weight %) was added to the biodiesel and stirred for ~25 minutes, then filtered off

After the pre-treatment, metathesis reactions of the purified reaction mixtures (self-metathesis) were carried out. The results showed that the pre-treatment did not lead to a significant improvement in results.

2.3.3 Microwave-assisted cross-metathesis

Microwave synthesis offers many advantages such as reduced reaction times and often improved reproducibility.

To accelerate fuel production, the cross-metathesis reaction of biodiesel was tested with 1-hexene in a microwave reactor. Two reactions were carried out using a biodiesel-hexene ratio of 1:0.8 with 0.05 mol% Umicore M5$_1$. One reaction was carried out with conventional heating and delayed catalyst addition (via pump) for four hours, whilst the other was carried out in a microwave reactor for only five minutes with the catalyst added immediately.

Catalyst	Temp [^{o}C]		Conversion [%]	TON
Umicore M5$_1$	60	Microwave	83.9	1678
		Standard	97.1	1942
	120	Microwave	86.7	1734
		Standard	95.1	1902

Table 2-3:   Comparison of reactions with delayed addition (pump) and a microwave reaction

The results in Table 2-3 tellingly show that after just five minutes microwave synthesis very high conversions of over 80% and TON figures could be achieved, which means a clear time advantage over the previously-used reaction time of four hours.

2.4   Catalyst removal

In order to enable engine tests of the metathesis fuel, the ruthenium catalyst must be removed after the metathesis reaction. Three different methods were tested: activated charcoal, silica gel and hydrogen peroxide.

Silica gel

With this method, the cross-metathesis mixtures were filtered through a column with a small quantity of silica gel directly after the reaction. This provides minimal removal of catalyst material. However, there was also a fuel loss, since this was in part absorbed by the silica gel. Since the purely visual examination did not show any significant reduction in catalyst quantity, this very simple method was not tested further.

Activated charcoal

The second tested method used activated charcoal (Darco, Aldrich), which was added to the cross-metathesis mixtures. After stirring for four hours, the mixture was filtered using a filter paper. The filter paper was not able to effectively retain the catalyst and the charcoal and hence dark solutions were obtained. This simple method also failed to yield the desired results.

Hydrogen peroxide

The third method for catalyst removal used hydrogen peroxide. Water, followed by aqueous hydrogen peroxide (35%) was added to the metathesis mixture and this mixture was stirred vigorously for about an hour. The molar ratio was 5800:1 hydrogen peroxide to ruthenium catalyst. This mixture was then separated in a separating funnel. The aqueous phase was extracted with hexane (2.14 ml) and the combined organic solutions were tested for peroxide (using test strips). The organic solution was then dried with Na_2SO_4 and filtered over a small quantity of silica gel. Then the organic solvent was separated via distillation.

The ruthenium contamination of the purified biodiesel was analysed via ICP-MS and the results showed that about 99% ruthenium had been removed. In an ideal case, the residual ruthenium content was just 6 ppm. This method is documented in Fig. 2-8 with pictures and also shows visually that this procedure successfully removes ruthenium. The obtained metathesis fuel even had a lighter colour than the original biodiesel.

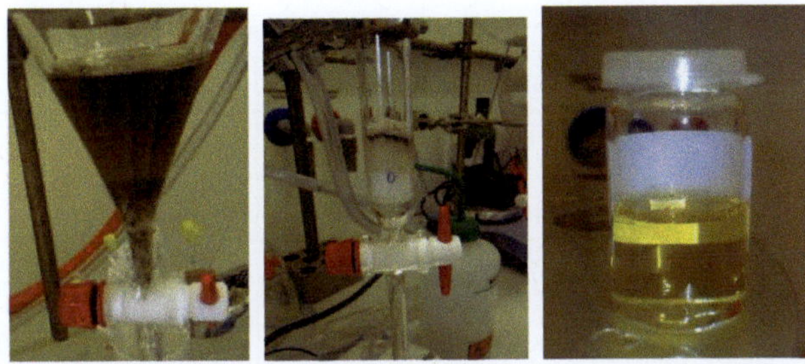

Fig. 2-8:   Formation of a dark precipitate on the underside of the separating funnel (left); catalyst residue on silica gel (centre); catalyst-free biodiesel (right)

2.5   Synthesis on a larger scale

Table 2-4 shows a list with some of the samples that were synthesised on a larger scale during the course of this project for engine tests. These samples were produced in a Radleys reactor which is connected with a thermostat.

Reaction	Volume [ml]	Umicore M5$_1$ [mol %]	Conversion [%]	TON	Note
A	1553	0.05	96.6	1932	Cat. added in 2 parts
B	1269	0.05	95.5	1909	Cat. added in 2 parts
C	2000	0.05	74.4	1487	Cat. added all at once
D	2000	0.05	71.8	1435	Cat. added all at once
E	2000	0.05	96.5	1931	Cat. added in 2 parts
F	2000	0.05	96.6	1931	Cat. added in 2 parts
G	2000	0.05	96.5	1930	Cat. added in 2 parts
H	2000	0.05	96.4	1928	Cat. added in 2 parts
I	2000	0.05	96.1	1922	Cat. added in 2 parts
J	2000	0.04	94.2	2355	Cat. added in 4 parts
K	2000	0.03	84.3	2808	Cat. added in 4 parts
L	2000	0.03	83.3	2777	Cat. added in 4 parts

Table 2-4:   Conversion and TON results of biodiesel synthesis on a larger scale

The presented results are derived from cross-metathesis reactions with a mixture of 1 eq biodiesel and 0.8 eq 1-hexene. The results correlate very clearly with the results of the small test reactions, since slow catalyst addition also yielded better conversions and TON here. Even with 0.03 mol% catalyst, conversion rates greater than 80% were attained.

2.6 Metathesis fuels for more extensive tests

All fuels synthesised at the Karlsruhe Institute for Technology for more extensive tests are listed again in Table 2-5 with their key properties.

Sample number	Cleaning	Description	Catalyst	Catalyst quantity
MA	no	not available	Grubbs 1st generation	n/a
MB	no	60% cross / 40% self-metathesis	Umicore M2	n/a
MC	no	71% cross / 29% self-metathesis	Umicore M2	n/a
MD	yes	1 eq:1 eq RME:1-hexene	Umicore M3$_1$	0.20% eq
ME	yes	Self-metathesis (pure RME)	n/a	0.10% eq
MF	yes	1 eq:1 eq RME:1-hexene	Umicore M5$_1$	0.05 mol%
MG	yes	1 eq:0.5 eq RME:1-hexene	Umicore M5$_1$	0.05 mol%
MH	yes	1 eq:0.2 eq RME:1-hexene	Umicore M5$_1$	0.05 mol%
ML	yes	1 eq:0.4 eq RME:1-hexene	Umicore M5$_1$	0.05 mol%
MJ	yes	1 eq:0.6 eq RME:1-hexene	Umicore M5$_1$	0.05 mol%
MK	yes	1 eq:0.8 eq RME:1-hexene	Umicore M5$_1$	0.05 mol%
ML	yes	Self-metathesis (pure RME)	Umicore M5$_1$	0.017 mol%
20-litre batches for commercial vehicle engine				
MM	yes	Self-metathesis (pure RME)	Umicore M5$_1$	0.02 mol%
MN	yes	1 eq:0.8 eq RME:1-hexene	Umicore M5$_1$	0.05 mol%

Table 2-5: utilised metathesis products with key manufacturing parameters

From fuel MD, the steps for purification of the products described in 2.4 took place for use in the combustion engine. In total, the project includes 14 metathesis products. The first four products were only used for preliminary tests and not for emissions measurements on the engine. Samples MM and MN were produced on a larger scale for tests on the commercial vehicle engine and almost completely correspond with the reaction conditions of products ML and MK.

All fuels were used in the engine tests as a 20% blend with 80% reference diesel fuel CEC RF-06-03 and were mixed at the vTI to this end. The aforementioned diesel fuel, rapeseed oil methyl ester (which also served for production of the metathesis products) and a B20 blend of DF and RME were used as comparison fuels.

Metathesis product ML is a further-developed form of self-metathesis (product ME). Hence the basic fuel tests were repeated with this product shortly before production of the 20-litre batches and correspondingly did not take place in a test series with the remaining samples.

3 Investigation of interactions and engine tests

In the following text, the materials and equipment used at the vTI will be described in greater detail and further background information will be provided on the determination and effects of the individual exhaust gas components. In addition, the test bench of the Institute for Internal Combustion

Engines of the TU Braunschweig (ivb), on which a test series was also carried out, will be presented.

3.1 Fuel properties investigation

Since the biodiesel altered via metathesis is a completely new kind of fuel, some fundamental properties required testing before it was used in engines. In particular, these included its behaviour with other operating materials and its material compatibility. In addition, the boiling curves of the produced metathesis fuels are of particular interest since altering the boiling curve was a main component of the project. The methods for determining these properties will be described in greater detail in the following text.

3.1.1 GC-FID analysis

A gas chromatograph with flame ionisation detector Star 3600 CX from the company Varian was used for the evaluation of the composition of biodiesel in comparison with the metathesis products. The technical data for the capillary column of the gas chromatograph and the utilised methods are specified in Table 3-1. Since the system was used without comparison with a standard, only comparative statements on the utilised and produced fuels can be made. It was not possible to determine individual components and their precise concentration in the fuel.

Manufacturer	Restek
Stationary phase	Crossbond 100% dimethyl polysiloxane
Polarity	strongly non-polar
Column length	10 m
Interior diameter	0.53 mm
Film thickness	2.65 μm
Temperature range	-60 to 400 °C
Temperature gradient	50 °C (3 min) - 15 °C/min - 300 °C (5 min)
Split ratio	Splitless mode
Carrier gas flow (at 50 °C)	12.8 mL/min
Injector temperature	300 °C isotherm
Detector temperature	300 °C isotherm
Injection volume	1 μL

Table 3-1:   Details of the capillary column, GC-FID and method

3.1.2 Boiling curves

The boiling curves of the individual fuels were determined using simulated distillation in accordance with ASTM D2887 using a gas chromatograph with flame ionisation detector 7890A from the company Agilent at the Automotive Technology Centre of the University of Applied Sciences and Arts, Coburg. The measuring device was calibrated with the boiling standard Polywax 1000.

This measurement process very quickly allowed precise statements to be made about the boiling behaviour of a fuel. The most important equipment data of the measurement system can be found in Table 3-2.

Column	Agilent J&W DB* HT SimDis
Length	5 m
Carrier gas flow	14 mL/min
Temperature gradient	40 °C - 20 °C/min - 350 °C
Injector temperature	350 °C
Split	10:1
Injection volume	0.5 µL
Detector temperature	350 °C
Combustion gas flow	40 mL/min (H_2) 450 mL/min (air)
Purge gas flow	40 mL/min (He)

Table 3-2:   Equipment data for the simulated distillation

3.1.3 Miscibility with other fuels and engine oil

Since the metathesis fuel in the engine tests was not to be utilised as a pure fuel but rather as a blend with 80% diesel fuel, miscibility of these two components had to be guaranteed. Here, miscibility means the qualitative assertion that upon mixing at least two different fluids, these mix completely to form a single homogeneous phase. The polarity of two fluids has great influence on their miscibility. Polar fluids mix well with other polar fluids and non-polar fluids mix well with other non-polar fluids. However, there is no absolute gauge for the polarity of substances, hence a prediction of miscibility is only theoretically possible in accordance with the following basic principle:
Miscibility is better with increasing similarity of the reciprocal forces between the particles of one component and another component.
In the miscibility tests, in addition to the two aforementioned components, the two market-relevant components RME and HVO and also aged RME were used. Alcohol was also used to dissolve precipitates.
In a further test series, a test was carried out on the behaviour of metathesis fuel in engine oil, since a continuous incursion of small quantities of fuel into the engine oil occurs during engine operation. This effect was tested by mixing engine oil (super low-friction oil DIMO Premium, SAE 10W-40, Karl H. Heusmann KG) with the metathesis fuel (MA). This oil-fuel mixture was then stored under different conditions to allow the identification of interactions. The samples were stored at -18 °C and room temperature for 24 hours and at 90 °C for 22.5 hours. Sampling at 90 °C was due to the temperature profile represented in Fig. 3-1. It comprised ten heating phases of 120 minutes respectively followed by cooling times of 15 minutes.
These conditions should indicate the temperature change occurring in the oil during engine operation. After storage, the mixtures were visually examined.

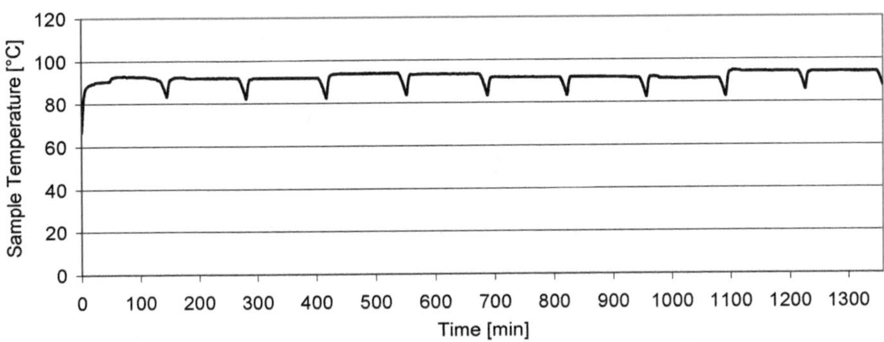

Fig. 3-1:   Temperature curve of heating

3.1.4 **Material compatibility**

For a new fuel, compatibility with the different materials with which it comes into contact is decisive. In order to be able to make an initial assessment of the material behaviour, two different material samples were tested with respect to their resistance to the fuel in line with DIN EN ISO 175.

To this end, the samples were stored for seven days at a temperature of 70 °C in 70 ml of the metathesis fuel (ML). To guarantee constant temperature conditions, the sample vessels were placed together in a water bath and the temperature was monitored with a thermometer in a further test tube with DF. Storage in fossil diesel fuel and rapeseed oil methyl ester provided comparison figures. In addition, reference samples were stored at 22 °C and 45% air humidity in a climatic chamber. Tensile test samples of type 1A in accordance with DIN EN ISO 527-2 made from polyamide (PA 66 Ultramid A3K) and high density polyethylene (HDPE Lupolen 4261) were used as test specimens. Material selection took place in consultation with an OEM and the Federal Institute for Materials Research (BAM). Both materials are classified as resistant to biodiesel. Before commencing storage, the individual samples were each weighed three times to determine their original mass. After storage, the mass change and visual inspection of the samples took place. Following this, the modulus of elasticity and tensile strength were determined in accordance with DIN EN ISO 527-2. All tests were carried out as triple determinations.

3.2 **Test engines**

Two different engines were used to test the emissions of the fuels altered by metathesis. The fuels that were produced in small quantities were first compared using the single-cylinder engine to select two advantageous fuels for tests using the commercial vehicle engine. The two utilised test engines will be presented in greater detail in the following text.

3.2.1 **Farymann single-cylinder test engine**

A single-cylinder engine made by Farymann was used to test the emissions of different metathesis products. It generated a maximum output of 5.2 kW. Peak engine speed was 3600 rpm and maximum torque was 15.3 Nm.

3.2.2 **Engine OM 904 LA**

An OM 904 LA commercial vehicle engine from Daimler AG was available for the measurement of selected fuels. The basic engine data are listed in Table 3-3.

Cylinder stroke	130 mm
Cylinder bore	102 mm
No. of cylinders	4
Capacity	4250 cc
Nominal speed	2200 rpm
Rated power	130 kW
Maximum torque	675 Nm at 1200 to 1600 rpm
Exhaust gas after-treatment	SCR catalytic converter
Exhaust gas standard	Euro IV

Table 3-3:   Technical data of the OM 904 LA test engine

The engine was fitted with an SCR catalytic converter for nitrogen oxide reduction. The engine model as installed on the test bench achieved the relevant boundary values in accordance with Euro IV. In series production, this engine is used in the Atego trucks of Daimler AG. For operation on the test bench, the engine was equipped with the standard production radiator and other ancillary parts. The engine was connected via a universal drive shaft with an asynchronous motor from the company AVL, via which both static and also dynamic test cycles can be realised. The system as a whole is shown in Fig. 3-2.

Fig. 3-2:   OM 904 LA engine installed on the test bench

The SCR catalytic converter can be seen marked with an arrow. The exhaust pipe section from the engine to the catalytic converter also comprised original components from the manufacturer. Determination of the emissions took place at the outlet of the catalytic converter. SCR catalytic converters for after-treatment of exhaust gases have been used for many years with stationary engines and in power stations. They were first used for mobile applications upon introduction of the Euro IV standard. Ammonia is used to reduce nitrogen oxide. Hence a second operating substance must be carried in a second tank in addition to the fuel. In addition to the tank, there are a few other additional components. The 32.5% aqueous urea solution (AdBlue) which serves to provide the ammonia is injected from the tank directly into the exhaust gas stream via a dosage system. After this, conversion to ammonia and reduction of the nitrogen oxides takes place in the catalytic converter.

3.2.3 Test cycles

The single-cylinder engine was operated in a five-point test to determine the emissions. The load points were selected in line with the five-point test for agricultural tractors (Welschof, 1981). These proved expedient in preliminary tests as they nicely cover the entire load range of the engine. In addition, they allowed the points listed in Table 3-4 to be kept constant over a longer period because here the engine exhibited only very slight drift towards other engine speeds and torques. Hence a high level of reproducibility was to be expected with these settings.

Operating point	Engine speed [rpm]	Torque [Nm]	Weighting factor
1	2840	10	0.31
2	2650	6	0.18
3	1850	4	0.19
4	3000	2	0.20
5	1050	0	0.12

Table 3-4:   Operating points for the single-cylinder test engine

The OM 904 LA commercial vehicle engine was operated in accordance with European Union guideline 2005/55/EC in European transient cycle (ETC). The test cycle is to be adapted to the full load curve of the engine. Determination of the emissions took place during the entire 30 minutes of the test. The cycle is divided into three parts. The first ten minutes reconstruct an urban driving part, the second ten minutes reconstruct a rual driving part and the third ten minutes reconstruct a motorway part. The engine speed and torque profile of the test can be found in Fig. 3-3. The specification of the load values took place in one-second intervals. This produced 1800 different load states over the entire course of the test.

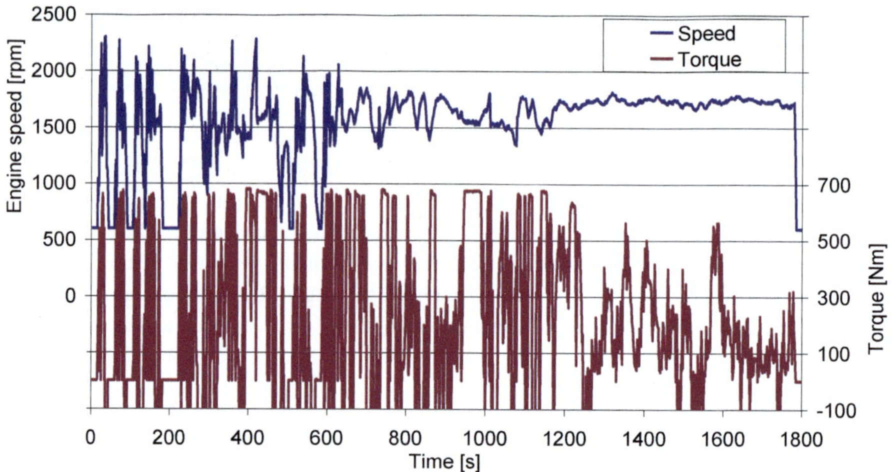

Fig. 3-3:   Engine speed and torque curve in ETC

3.2.4 AVL single-cylinder test engine based on MAN D28

In order to test the suitability of the fuels for use in engines with respect to their emissions and combustion behaviour, measurements were carried out on two blends of fossil diesel fuel and different metathesis fuels on a single-cylinder commercial vehicle test engine at the Institute of Internal Combustion Engines at the University of Braunschweig. A CEC reference diesel fuel, a CEC reference diesel fuel provided by vTI and a B20 blend with conventional RME served as comparison fuels for the tests. The evaluation comprised measurements of the statutorily regulated gaseous emissions, determination of the particle emissions via smoke meter and SMPS and also observation of the combustion behaviour via thermodynamic analysis.

The tests were carried out on an AVL single-cylinder test engine based on the MAN D28. The test engine was equipped with four valves per cylinder and a 2nd generation Bosch common rail injection system. In addition, the engine had external charging via an electrically-driven worm compressor, a flow heater for induction air conditioning and temperature-controlled external exhaust gas recirculation (EGR). Some engine data are summarised in Table 3-5.

Stroke	160 mm
Bore	128 mm
Capacity	2059 cc
Connecting rod length	251 mm
Ignition pressure threshold	220 bar
No. of valves	4
Combustion chamber shape	ω type
Compression ratio	14:1 or 16.5:1 – depending on piston type

Table 3-5:   Engine data for the AVL single-cylinder test engine

Engine control was handled by the institute's own engine control unit based on the C167 processor, which allowed independent influencing of the various control and regulation variables. The C167 engine control unit was supplemented by a freely-parametrisable industrial controller from the company Bürkert. In addition to the conventional measurement process for determination of engine speed, torque, fuel consumption and temperatures, the pressure curve of the cylinder was also recorded. The inlet manifold pressure and exhaust pressure of the cylinder were also recorded, resolved for crank angle.

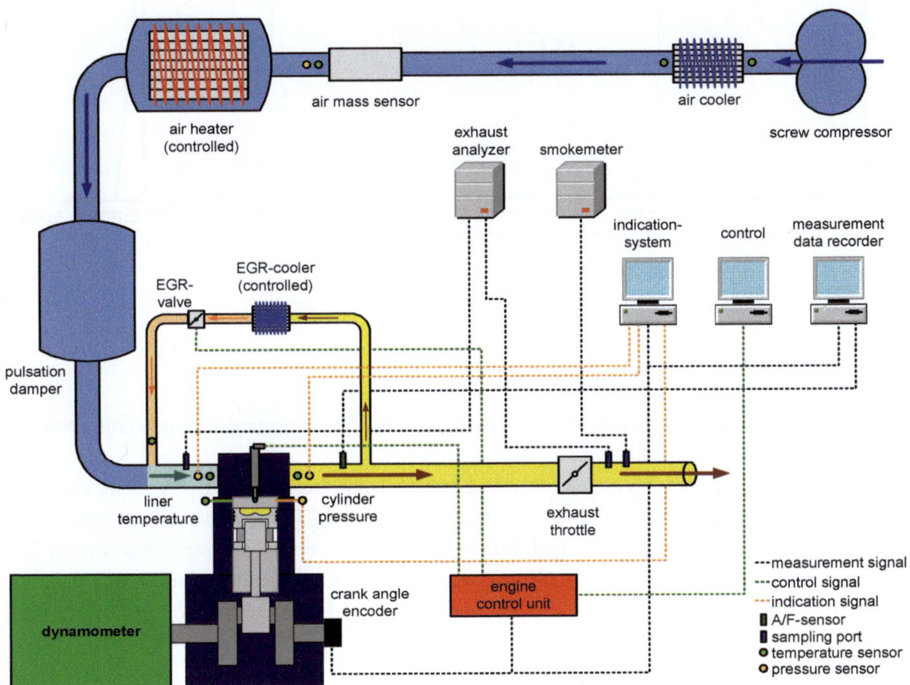

Fig. 3-4:   Configuration of the test station, schematic

Measurement of the regulated gaseous exhaust gas components took place using an EXSA 1500 analysis unit from the company Horiba; particle emissions were tested using an AVL smoke meter of type 415S and an SMPS from the company TSI. Fig. 3-4 shows a schematic representation of the configuration of the test bench.

Since the available test apparatus is a single-cylinder engine based on a commercial vehicle engine, the operating points are derived from the ESC test cycle. The cycle is represented in Fig. 3-5.

Since not all operating points of the text cycle could be covered due to the restricted availability of the metathesis fuels, a selection of five operating points were selected for implementation of the tests. The test matrix is composed of load points 1, 3, 5, 7 and 9 of the presented cycle.

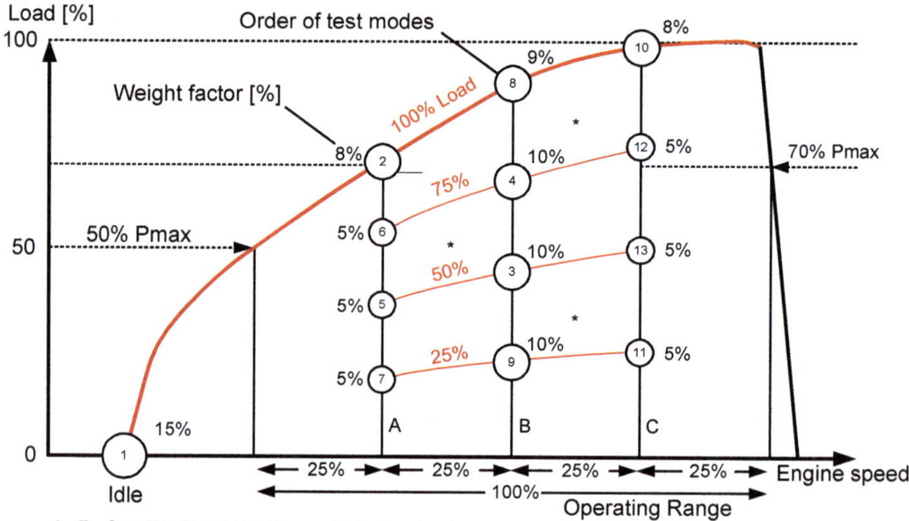

Fig. 3-5:   ESC test cycle

After adaptation to the specifications of the test apparatus, the measurement programme presented in Table 3-6 was produced from this.

The setting parameters are orientated towards a standard full engine with EU V EGR emissions concept with open particle filter system and without SCR system. The engine load was regulated to an indicated medium pressure of the high-pressure loop, since in this way any deviations in load change do not have any influence on the measurements. Therefore calculation of the specific emissions is also carried out based on the indicated load without load change losses.

Operating point	Speed [rpm]	Load $p_{mi,HD}$ [bar]	λ_{Start} without EGR [-]	EGR rate [%]	HV 50 [°CA ATDC]	Rail pressure [bar]
1	1100	6	3.5	0	5	1200
2	1100	6		25	5	1200
3	1100	12	2.5	0	10	1400
4	1100	12		25	10	1400
5	1450	6	3.5	0	5	1200
6	1450	6		30	5	1200
7	1450	12	2.5	0	10	1400
8	1450	12		25	10	1400
9	850	2	6.5	0	1	1000

Table 3-6:   Measurement programme for the single-cylinder test engine

The fuel parameters required according to Brettschneider for determination of the combustion air ratio via the function stored in the exhaust gas analysis unit, such as molar H/C ratio, molar O/C ratio and the stoichiometric minimum air requirements, can be calculated from the analysis results. A set of these parameters is presented in Table 3-7.

Parameter	CEC DK	DK 9 vTI	B20 vTI	MM20 vTI	MN20 vTI
C portion [% m/m]	86.36	86.41	84.53	84.53	84.61
H portion [% m/m]	13.64	13.59	13.27	13.27	13.33
O portion [% m/m]	0.00	0.00	2.00	2.10	1.96
H/C mol calc	1.882	1.874	1.871	1.871	1.877
O/C mol calc	0.000	0.000	0.018	0.019	0.017
Air-fuel ratio$_{stoich}$	14.57	14.56	14.18	14.16	14.19

Table 3-7:   Fuel parameters

3.3 Analysis of exhaust gas emissions of Farymann and OM 904 LA

The analysis of exhaust gas components is divided into two areas. On the one hand, the statutorily regulated emissions are considered. These include carbon monoxide (CO), total hydrocarbons (HC), nitrogen oxides (NO_x) and particulate matter (PM).

The second area contains the non-regulated emissions. Here, particle size distribution (PSD), ammonia slip (NH_3), polycyclic aromatic hydrocarbons (PAH) and carbonyls were tested. This section will examine the individual components and the associated analysis process in greater detail. The statements relating to the effects of the pollutants on humans and the environment generally correspond with the dissertation by Stein (2008) produced at the institute.

3.3.1 **Carbon monoxide (CO)**

Carbon monoxide is formed during the combustion in a diesel engine in areas with a lack of oxygen. For the most part, however, the CO is converted to carbon dioxide by post-oxidation before it even leaves the cylinder (van Basshuysen and Schäfer, 2002).

CO is an odourless and colourless gas that collects on the ground as it is somewhat heavier than air. The toxic effect of carbon monoxide emissions arises through the strong desire of the CO molecules to permanently bond with haemoglobin in the blood, which then leads to problems with oxygen transportation. Hence the emissions should be kept as low as possible. According to a calculation model commissioned by the German Automobile Industry Association (VDA) and the Federal Environmental Agency, a reduction of 90% is estimated for the period from 1990 to 2020 in road traffic (VDA, 2010).

The CO emissions are determined using a gas analyser from the company Bühler Technologies. The sample gas stream is fed through two cuvettes, divided into two equal partial streams. One of the two cuvettes is illuminated with non-dispersed infrared light, the wavelength of which is tuned to the absorption of carbon monoxide. The temperature and pressure increase that is generated by the energy supplied in this way leads to a compensation flow between the two chambers. This compensation flow can be determined with the help of a microflow sensor, which is a measure of the CO concentration in the cuvettes.

3.3.2 **Hydrocarbons (HC)**

In contrast to carbon monoxide, hydrocarbons in some cases cause considerable eco-toxicological effects. Primarily, these contribute towards photochemical oxidant formation ("summer smog"). Due to photochemical reaction of the hydrocarbons with nitrogen oxides, ground-level ozone as well as aldehydes and ketones among other things are produced (Mollenhauer, 2002). These substances can contribute towards acute and chronic damage to the respiratory tract and an increase in the risk of illnesses of the cardiovascular system.

Hydrocarbons in the diesel exhaust gas primarily result from incomplete combustion. This especially occurs in zones that are not involved in the combustion, in which there is a lack of air or in which flames are extinguished by too low temperatures, for example near the walls (Hackbarth und Merhof, 1998). When using SCR catalytic converters for exhaust gas after-treatment, barely any HC emissions are to be expected since the system converts the non-combusted fuel components with an efficiency in the range of >90% (Gekas, 2002).

A gas analyser RS 55-T from the company Ratfisch is used to identify the hydrocarbons. It works with a flame ionisation detector (FID). In the device, the sample gas is ionised in a helium-hydrogen flame that burns in an electric field. The hydrocarbon content is determined by measuring the field change. Calibration takes place with propane (C_3H_8, 91.5 ppm) as single-point calibration. In doing this, a pipe that is kept heated to 190 °C with the aid a thermostat supplies the hot and pre-filtered exhaust gas to the HC analyser. This heating of the gas flow is designed to prevent premature condensation of higher boiling hydrocarbons.

3.3.3 Nitrogen oxides (NO$_x$)

On combustion, nitrogen oxides are predominantly formed in oxygen-rich areas of the combustion chamber. In the literature, the Zeldovich reaction is described for their formation (Fernando et al., 2006; Warnatz et al., 2001).

The individual steps are described in equations 6 to 9.

$$O_2 \rightarrow 2O \cdot \qquad\qquad\qquad (6)$$

$$O \cdot + N_2 \rightarrow NO + N \cdot \qquad\qquad\qquad (7)$$

$$N \cdot + O_2 \rightarrow NO + O \cdot \qquad\qquad\qquad (8)$$

$$N \cdot + \cdot OH \rightarrow NO + H \cdot \qquad\qquad\qquad (9)$$

Since a very high level of activation energy in the form of high temperatures is required for the formation of nitrogen monoxide in accordance with equation (7), this is also known as "thermal NO". Due to its rapid conversion to NO$_2$ in the atmosphere, NO is toxicologically unimportant. However, nitrogen dioxide can penetrate very deeply into the respiratory tract and can form nitric and nitrous acids at the transition into the bloodstream, which can damage cells. The entire global road traffic network produces 17.5% of anthropogenic NO$_x$ emissions (Mollenhauer, 2002). In Germany, road traffic caused the majority of nitrogen oxide emissions in 2008 with a figure of 45%. From 1990 to 2008, total emissions reduced by 52% (UBA, 2010). Based on the German traffic sector, diesel cars and lorries were responsible for 9% and 53% respectively of traffic nitrogen oxide emissions in 2002 (VDA, 2004).

The nitrogen oxides are analysed using a chemiluminescence detector (CLD) from the company EcoPhysics (CLD 700 EL ht). The basis of the CLD measurement process is the fact that, upon oxidation of NO to NO$_2$, about 10% of the NO$_2$ molecules take on an electronically-excited state, from which they immediately return to a non-excited state, wherein photons are emitted (luminescence). These are identified and are a gauge for the NO content. To determine the entire content of NO$_x$ (NO + NO$_2$ = NO$_x$), the hot and filtered sample is first supplied through a converter in which NO$_2$ is reduced to NO. NO$_2$ is calculated as the difference from the measurement of NO$_x$ and NO. Using an ozone generator contained in the device, the CLD generates ozone (O$_3$) for the oxidation of NO to NO$_2$. A calibration gas with 197 ppm NO in nitrogen and a second calibration gas with 52 ppm NO$_2$ in synthetic air are used for calibration.

3.3.4 Particulate matter (PM)

There is no generally-accepted definition for diesel particles. According to the definition of the Environmental Protection Agency in the USA (EPA), particles mean all materials that are present in solid or fluid form at temperatures below 51.7 °C in diluted exhaust gas and can be precipitated onto a filter (Code of Federal Regulations). The temperature of 51.7 °C corresponds to the American specification of 125 °F. Due to the restricted temperature, higher boiling components of the exhaust gas are precipitated onto the filter in condensed or adsorbed form.

A main component of the particles is non-combusted carbon. This soot arises in areas of the combustion chamber with a high temperature, low oxygen proportion and an excess of fuel. The precise formation process has still not been completely explained. However, two possible reaction paths are widely recognised. One of the possibilities for soot formation is schematically represented in Fig. 3-6.

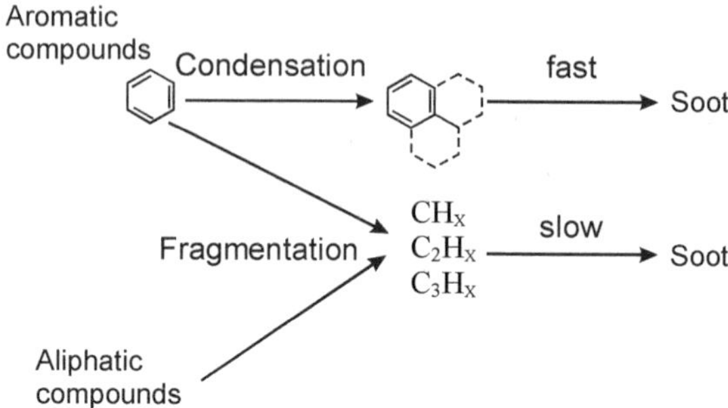

Fig. 3-6:   Schematic representation of soot formation   (according to Amman and Siegla, 1982)

Sampling particulate matter took place in an exhaust gas partial flow dilution tunnel (Fig. 3-7) which was set out in accordance with BS ISO 16183:2002 (2002).

In this dilution tunnel, the exhaust gas is diluted with purified compressed air and hence cooled to below 51.7 °C. The particles are collected on a two-stage PTFE filter by drawing the diluted exhaust gas through the filter. Extraction through the dilution tunnel is kept constant through a mass flow regulator. The supply of dilution air is determined by the mass flow regulator. The volume flow of exhaust gas that enters the dilution tunnel results from the difference of flows. It is regulated so that it always corresponds to a constant proportion of overall gas volume flow. Regulation of the dilution system takes place by determination of the dynamic pressure in the exhaust pipe and in the sampling probe.

The volume that is drawn through the filter results from the integral of the volume flows of the mass flow regulator over the course of the test. In this way, the particle mass in the entire engine exhaust gas can be inferred from the particle mass precipitated on the filter using the exhaust gas overall volume flow and the obtained exhaust gas proportion.

The partial flow dilution with full flow sampling is represented schematically in Fig. 3-7.

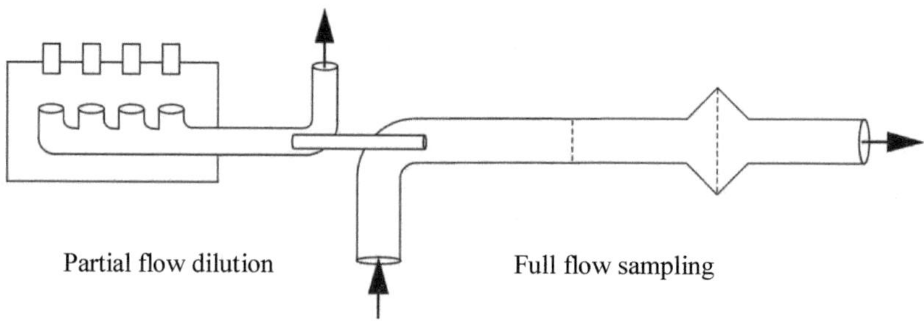

Fig. 3-7:   Schematic representation of the dilution system

3.3.5 **Particle size distribution**

Particle size distribution has recently gained in importance, as in future a threshold value for particle number is to be met in addition to the particulate matter threshold value (EC Regulation 595/2009). With this development, it is recognised that in particular the small particles enter far into the respiratory tract and can cause damage due to pollutants such as polycyclic aromatic hydrocarbons condensing on the surface.

Determination of the particle size distribution takes place with the help of an electric low pressure impactor (ELPI) from the company Dekati Ltd.

With this device, a size range from 28 nm to 10 µm aerodynamic diameter can be determined, divided into twelve levels. Precipitation takes place in an impactor based on the speed inertia ratio of the different particle fractions. To determine the number of precipitated particles for each level, the particles are charged by corona discharge with a voltage of 5 kV. When the charged particles impact on the impactor discs, a measurement voltage dependent on the quantity occurs.

3.3.6 **Ammonia**

Ammonia is a colourless, pungent-smelling and very water-soluble gas. In the atmosphere, NH_3 plays an important role as a base since it can neutralise acids such as H_2SO_4 and HNO_3. As this takes place, secondary aerosol particles are formed via ammonium salts that, for example, have an effect on cloud formation and human health (Baek et al., 2004). In the ground, it can cause acidification. NH_3 is also harmful to humans as a gas. Due to caustic properties, it can cause irritation to eyes and mucous membranes. The workplace threshold value for ammonia is 20 ppm (BMAS, 2006). Its high degree of water-solubility considerably enhances its penetration into the human body. Due to its low durability in the atmosphere, ammonia primarily has an effect in the area of its source of emission.

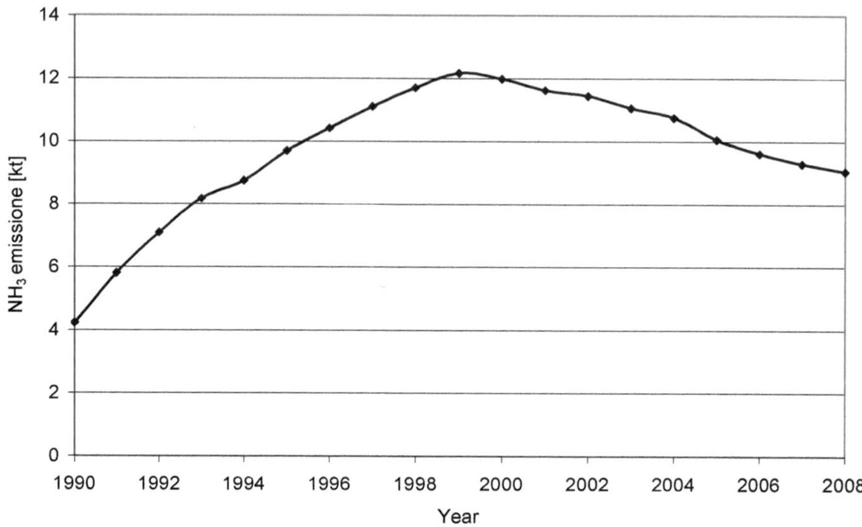

Fig. 3-8:   Development of traffic-related NH$_3$ emissions in Germany (according to UBA, 2010)

The main emission source of NH$_3$ is agriculture with approx. 95%. In Germany, traffic only contributes about 1.5% to overall ammonia emissions (UBA, 2010). The development of traffic-related ammonia emissions in Germany between 1990 and 2008 is presented in Fig. 3-8. The increase from 1990 to 2000 can be explained by the use of three-way catalysts. Up until 2008 emissions dropped again due to further developments in catalytic converter and control technology.

Estimates by the German Federal Environmental Agency assume that the NH$_3$ emissions will continue to drop to approx. five kilotons and will then remain constant until 2020 (UBA, 2003). However, a contribution from ammonia residues due to incomplete conversion during exhaust gas aftertreatment with SCR catalytic converters has hitherto not been taken into consideration. An increase in traffic-related emissions cannot be excluded due to the standardised use of these systems. A threshold of 10 ppm for NH$_3$ will be introduced with introduction of the Euro VI exhaust gas standard to prevent a strong increase in emissions.

Determination of the ammonia slip took place using a mass spectrometer with chemical ionisation (CI-MS) from the company V&F Analyse- und Messtechnik GmbH.

3.3.7 Carbonyls

High concentrations of diesel engine emissions cause acute irritation of the mucous membranes in the upper respiratory tract and eyes. This is primarily an effect of gaseous components in the exhaust gas (Scheepers and Bos, 1992), to which primarily the aldehydes belong in addition to nitrogen oxides. The highest concentrations are found for acetaldehyde, formaldehyde and acrolein (Nold and Bochmann, 1999).

Emissions of aldehydes and ketones have been determined with DNPH cartridges. These cartridges contain silica gel that is coated with 2,4-dinitrophenylhydrazine (DNPH). If aldehydes or ketones are conducted via the DNPH, the analogous hydrazones result in accordance with the reaction contained in Fig. 3-9.

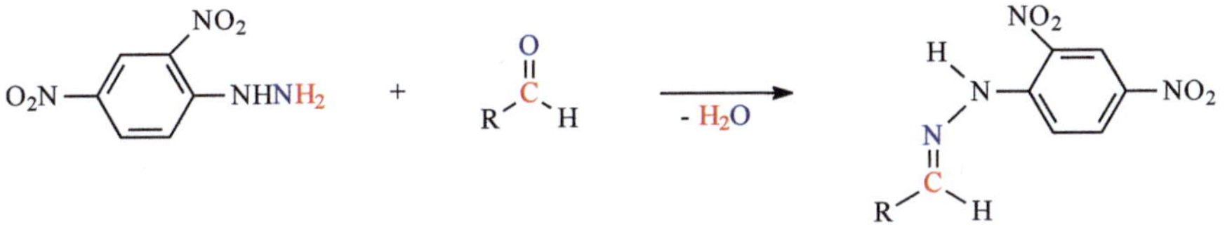

2,4-dinitrophenylhydrazine   **Aldehyde**   **Aldehyde**-2,4-dinitrophenylhydrazone

Fig. 3-9:   Exemplary evidence of an aldehyde by derivatisation to hydrazone

Sampling took place from the filtered raw exhaust gas. The exhaust gas was heated up to the cartridge so that no condensation occurred. Independent of the exhaust gas volume flow, the flow speed through the cartridges was constant at 0.5 l/min.

Since DNPH reacts with nitrogen dioxide (NO_2) from the exhaust gas, another cartridge with potassium iodide is included before the DNPH cartridges. The NO_2 is reduced by the iodine, wherein NO^{2+} can be precipitated in the cartridge and no longer influences the concentration of aldehydes and ketones in the DNPH cartridge. The formed hydrazones are rinsed with acetonitrile from the cartridges in a 2 ml volumetric flask.

An HPLC system from the company VWR was used to analyse the solution. Detection took place using a diode array detector.

Elution sequence	Analyte
1	Formaldehyde DNPH
2	Acetaldehyde DNPH
3	Acrolein DNPH
4	Acetone DNPH
5	Propionaldehyde DNPH
6	Crotonaldehyde DNPH
7	Methacrolein DNPH
8 + 9	2-butanone DNPH and n-butyraldehyde DNPH
10	Benzaldehyde DNPH
11	Valeraldehyde DNPH
12	m-tolualdehyde DNPH
13	Hexaldehyde DNPH

Table 3-8:   Individual substances of the carbonyl standard

The DAD spectrum serves for identification of substances. For the evaluation, the chromatogram was used at a wavelength of 370 nm. Calibration of the system takes place with a carbonyl standard with 13 individual substances (Cerilliant company) that are listed in Table 3-8. Due to the possible application as solvents in the laboratory, the respectively determined acetone concentrations were not specified in the framework of the project to prevent misunderstandings.

The technical data for the columns used are summarised in Table 3-9.

Manufacturer	Merck
Column length	250 mm
Interior diameter	4.6 mm
Grain size	5 μm
Column temperature	36 °C
Injection volume	10 μL
Mobile phase	Water / acetonitrile (gradient programme)
Flow	0.5 mL/min

Table 3-9:   Technical data of the HPLC column LiChrospher® 100 RP-18

3.3.8 Polycyclic aromatic hydrocarbons

One speaks of polycyclic aromatic hydrocarbons (PAH) if an organic compound comprises two or more aromatic rings. In accordance with the current standards, PAH may be present in quantities of up to 8% in diesel fuel (DIN EN 590, 2010). However, they are also formed during incomplete diesel engine combustion and represent a possible first step in the formation of soot particles (cf. Fig. 3-6). At this point in the formation, part of the PAH can convert to the gaseous phase and later adsorb on the soot particle surfaces. The precise way to form PAH is not definitively clarified.

PAH are the most common and well-known of the genotoxic or carcinogenic chemical compounds that are present in the ambient air (Savela et al., 2003). Due to their low volatility, the PAH are generally carried in dust – or, in the exhaust gas, soot. Hence their distribution depends predominantly on the transportation of these particles in air. Therefore, absorption of the PAH takes place via inhalation of the particles. In this way, the polycyclic aromatic hydrocarbons can gain access to the bronchioli and then the blood (GESTIS materials database, 2011). Here they can exhibit their damaging effect, for example in the form of lung cancer.

The sampling apparatus for determination of particle-bonded polycyclic aromatic hydrocarbons is structured in accordance with VDI Guideline 3872 Page 1 (1989). The particles are collected on PTFE-coated fibreglass filters (Pallflex Fiberfilm, T60A20, 70 mm, Pall company) from the undiluted raw exhaust gas in the course of the implemented test run. In addition, components precipitated from the gaseous phase were precipitated in a cooling system cooled to -15 °C and formed a condensate that was also tested. Extraction of the filter took place with toluene in a standard extractor fexIKA 50 (IKA company) and was carried out for four hours. After completion of the extraction and concentration of the sample in a rotary evaporator, the sample was recrystallised in acetonitrile.

Firstly, the internal standard para-quaterphenyl was added to the condensate before extraction. After transferring to a separating funnel, a 1:1 mixture of toluene:dichloromethane was added. If necessary, toluene was added until phase separation occurred. The sample was extracted in an ultrasonic bath and then shaken out. This process was repeated three times and the extract was transferred into a round-bottomed flask. The contents of the flask were then recrystallised in accordance with the process described above.

Name	Number of rings	Utilised abbreviation
Naphthalene	2	Nap
Acenaphthene	3	Ace
Fluorene	3	Flu
Phenanthrene	3	Phe
Anthracene	3	Ant
Fluoranthene	4	Fla
Pyrene	4	Pyr
Benz[a]anthracene	4	BaA
Chrysene	4	Chr
Benzo[b]fluoranthene	5	BbFla
Benzo[k]fluoranthene	5	BkFla
Benzo[a]pyrene	5	BaPyr
Dibenz[a,h]anthracene	5	DBAnt
Benzo[ghi]perylene	6	BPer
Indeno[1,2,3-cd]pyrene	6	IPyr
Acenaphthylene	3	non-fluorescent

Table 3-10:   List of PAH determined from the exhaust gas

The samples generated in this way were analysed with an HPLC with fluorescence detector (model L-2480, VWR/Hitachi company). Here, the 16 PAH that are classified as particularly relevant by the US Environmental Protection Agency were tested. The compounds are listed in Table 3-10. However, in the analysis only 15 of the 16 listed PAH could be taken into consideration since acenaphtylene is non-fluorescent and hence could not be measured by the utilised detector.

3.3.9 Mutagenicity

The high mutagenic potency of diesel particle extracts was first described by Huisingh et al. 1978 and was rapidly confirmed by many other workgroups (Clark and Vigil, 1980; Claxton and Barnes, 1981; Lewtas, 1983). Further tests also showed that the particles themselves (Brooks et al., 1980; Siak et al., 1981; Belisario et al., 1984) and the condensates of the gaseous phase of DEE are mutagenic in the Ames test (Stump et al., 1982; Rannug et al., 1983; Matsushita et al., 1986).
Mutagenicity tests by Bünger et al. (1998 and 2000) and Krahl et al. (2001 and 2003) in relation to this clearly pointed to a higher mutagenic potential of diesel particulate from mineral diesel fuels than from fuels based on fatty acid methyl esters.
The direct mutagenicity of diesel soot particles is attributed to substituted polycyclic aromatic hydrocarbons, primarily nitro-PAH (Wang et al., 1978; Pederson and Siak, 1981; Ohe, 1984). The

native PAH must be converted into active metabolites to have a mutagenic effect. In the Ames test, this metabolic activation was attained by rat liver microsomes whose enzyme system had been processed by pre-treatment of the rats with Arochlor or other enzyme-inducing substances (Clark and Vigil, 1980). In a series of tests, nitro-PAHs were identified as the principle causes of the in vitro genotoxicity of organic extracts of diesel engine exhaust gases(overview in: Rosenkranz and Mermelstein, 1983).

In human and animal cell lines, a genotoxic effect of DEE extracts using different end points (chromosome aberrations, sister chromatide exchange, point mutation, DNA strand breakage, DNA repair) was also ensured (overviews at: German Research Foundation, 1987; Shirnamé-Moré, 1995).

The procedure for testing the mutagenicity of the particles was analogous to sampling the PAHs. Gaseous components were separated using an intensive cooler at -18 °C as a condensate. Separation of the components adsorbed by the glass wall took place by rinsing with 100 mL methanol. The coated filter and the condensate flask were stored at -20 °C and transported frozen to the Institute for Preventative and Occupational Medicine of the German Statutory Accident Insurance Scheme at the Ruhr University, Bochum (IPA). The extractions and the further preparation of the samples, as well as the test for mutagenic effects, were carried out here.

The particulates collected on the filter were subjected to Soxhlet extraction with 150 mL dichloro-methane in the dark (Claxton, 1983). This process enables the most effective production of mutagens from diesel exhaust gas particulates (Siak et al., 1981). The obtained extracts and condensates were reduced in a rotary evaporator and further dried under a nitrogen stream. For the Ames test, the dried extract was dissolved in 4 ml DMSO (Bünger et al., 1998).

The so-called Ames test (Ames et al., 1973 and 1975) covers the mutagenic properties of a wide spectrum of chemical substances and mixtures through the reverse mutation of a series of different test strains. These carry mutations in histidine operon. The mutations cause histidine auxotrophy in the test strains in contrast to the naturally-occurring strains of *Salmonella typhimurium*, which are histidine-prototrophic. The Ames test is the world's most widely used in vitro test process for testing the mutagenicity of complex mixtures such as combustion products and since 1997 has been recognised by the OECD as Guideline 471 "Bacterial Reverse Mutation Test". This study uses the revised standard test protocol of Maron and Ames from 1983 with test strains TA 98 and TA 100. TA 98 detects frameshift mutations and TA 100 base pair substitutions. The test strains were kindly provided by Prof. B.N. Ames.

The tests were carried out with and without metabolic activation via microsomal monooxygenases (S9 fraction). To this end, S9 induced with phenobarbital and benzoflavon from the company Tri-nova Biochem, Gießen, LOT no. 2427, manufactured on 01.06.2009, was used. The S9 mix was produced in accordance with the instructions from Maron and Ames (1983). The mutagenic methyl methanesulphonate (MMS), 2-aminofluorene (2-AF) and 3-nitrobenzanthrone (3-NBA) were used as positive controls.

The extracts were dissolved in 4 ml DMSO directly before the test process. A decreasing dilution series was also produced using DMSO, and this was used for the tests. 2-AF (100 µg/mL) and 3-NBA (10 ng/mL) were also dissolved in DMSO. MMS was diluted with distilled water (10 µg/mL). Fluid top agar (2.5 mL), which contained 0.05 mmol histidine and 0.05 mmol biotin, was mixed with 100 µL of a test concentration of the extracts and 100 µL of an overnight culture of a test strain. After shaking for a short time on a vortex, the mixture was spread directly on a minimal agar plate containing Vogel-Bonner medium E. Each test concentration was tested with both test strains and with and without the addition of 4% S9. Each extract was tested at least three times.

The colony count of reverse mutations on the Petri dish were counted after incubating for 48 hours at 37 °C in the dark. The background growth of the bacteria was regularly checked via light microscopy, since high concentrations of the extracts have a toxic effect on the test strains and lead to dilution of the background and a reversal of the mutations. The plates were counted with the help of an electronic colony counter (Biocount 5000 Pro A, Biosys, Karben, Germany). Routinely, 10% of the plates were counted manually as a control. The dose-effect curves of the Ames test typically show an initial linear increase in mutations. Due to increasing toxicity, the increase reduces at higher concentrations and can fall again in the event of strong toxic effects. The increase of the initial linear part of the curve is used as a gauge for the strength of the mutagenicity (Krewski et al., 1992). In these tests, the results were deemed positive if the colony count of the reverse mutations on the Petri dishes exhibited a dose-dependent, reproducible increase (Mortelmans and Zeiger, 2000). As standard, the significance of the increase in the dose-effect curves was tested with linear regression.

4 Results

The following section shows all results of the tests that were carried out with the produced metathesis fuels. First to be carried out were tests on fuel properties and their behaviour with respect to materials and other operating substances. After these results became available, emissions measurements could be carried out on the single-cylinder engine, which formed a prerequisite for decisions on the subsequent procedure. From these results, tests of two selected fuels on the OM 904 LA commercial vehicle engine were carried out.

4.1 Investigation of fuel properties

The new fuels produced in the project were tested from different viewpoints. The first analyses took place with metathesis fuels whose production did not yet contain any purification steps. In the further course, a few purification steps took place to purify the fuels for better compatibility with the engines used for emissions measurements.

The basic effects of the metathesis reaction can clearly be seen in Fig. 4-1. The green trace represents the measurement of RME with the contained components (C16, C18 and C20). Here, only a few signals can be seen for a retention time between 750 s and 1050 s. Each of these peaks illustrates a molecule contained in the biodiesel. A completely different picture is shown for the two measured metathesis products. There is a much higher number of peaks here, many of which were not present in the biodiesel. These signals show the new molecules that resulted from the metathesis reaction and that lead to a change in the boiling curve.

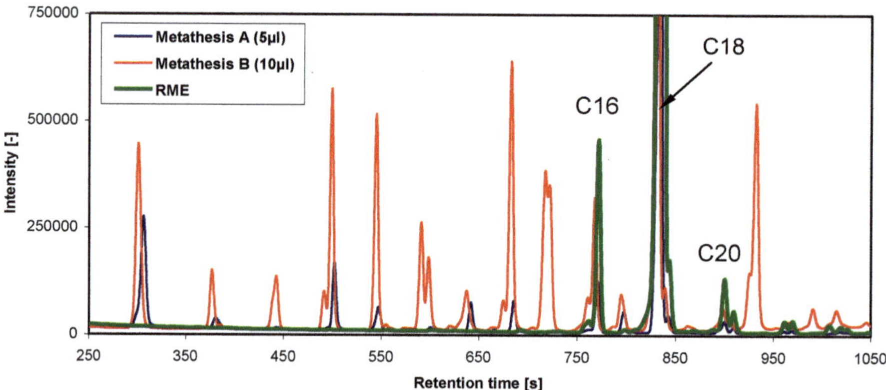

Fig. 4-1:   GC-FID spectrum of two metathesis products in comparison with RME

Furthermore, it can be seen that the two measured metathesis products differ from one another and some products are apparently only formed with one of the applied reaction conditions. However, to a certain extent the substances contained in the RME are still present after the reaction or have been reformed in the course of the reaction, which can be seen in the signals obtained for all three fuels.

4.1.1 Boiling curves

A main goal of the project is alignment of the boiling behaviour of biodiesel to diesel fuel. Hence the first step in analysis of the different metathesis fuels is determination of the boiling curve. The boiling curves of the first three metathesis products are shown in Fig. 4-2.

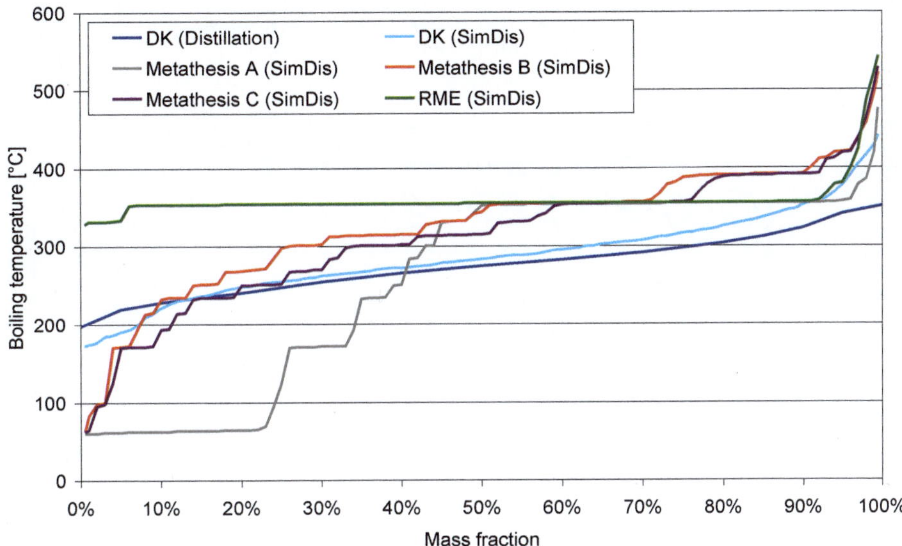

Fig. 4-2:   Boiling curves of the metathesis fuels without purification

With these three components, no purification was carried out after the metathesis reaction. By way of comparison, the boiling behaviour of fossil diesel fuel and rapeseed oil methyl ester are also listed. The boiling curve of RME is based on the same fuel that was used for the metathesis reaction. For DF, at the same time the course of a real distillation is represented that confirms the comparability of the simulated process with the real boiling curve by the minimal differences between the two curves.

It is clear from the graphic that the boiling curve of biodiesel can be strongly influenced by the applied metathesis reactions. In metathesis product A, approx. 40% of the components boil at lower temperatures than with fossil diesel fuel. The first 20% of the components even boil below 100 °C and therefore clearly lie below the desired boiling curve of the fossil diesel fuel. For this reason, the boiling curves of metathesis products B and C are clearly preferable. It becomes clear that the newly produced fuel only attains the boiling point of the starting material biodiesel at approx. 50%. In this way, problems such as oil dilution that result from biodiesel entering the engine oil (Tschöke et al., 2008) could be substantially reduced. Due to the lower boiling point, 40 to 50% of the entered fuel could leave the engine oil again. In this way, correspondingly longer oil change intervals are possible, based on the use of biodiesel.

As already mentioned under 2.6, purification of further products took place after the metathesis reaction to remove the catalyst. In addition, the ratios of biodiesel to the reaction partner hexene were varied. The precise conditions for production of the individual fuels are explained in chapter 2. The boiling curves presented in Fig. 4-3 are produced for the purified products.

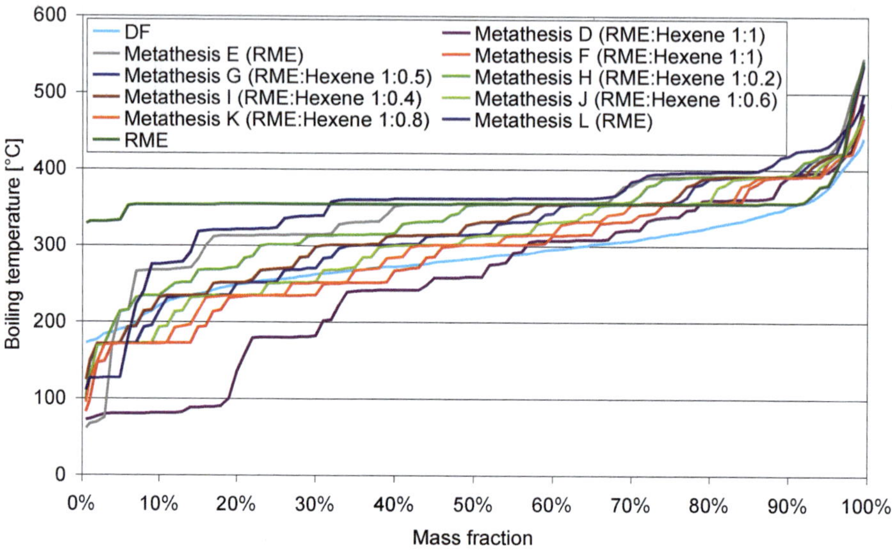

Fig. 4-3:   Boiling curves of the metathesis products obtained with different 1-hexene proportions after purification

As was to be expected, shorter molecule chains that lead to lower boiling temperatures result when the hexene proportion is increased. Furthermore, it is clear that the resultant boiling point depends on the type and quantity of utilised catalyst with the same ratio of biodiesel and 1-hexene. Hence samples D and F exhibit a clearly different boiling curve with the same ratio. For sample D with the lower boiling point, 0.2% of the catalyst Umicore $M3_1$ was used, for sample F 0.05% of the catalyst Umicore $M5_1$ was used. Furthermore it is obvious that when using self-metathesis without hexene, the boiling curve already attains that of the biodiesel in the range from 30 to 40% and due to the longer chains that also result, first exceeds the boiling point of the RME. This behaviour is represented by the products E and L in Fig. 4-3. With a high 1-hexene proportion of 1:1, the boiling curve only attains the RME value at approx. 75%.

Since the fuels in the engine tests were used as blends with 20% metathesis fuel in fossil diesel fuel, simulated distillation was also carried out with some of these mixtures to test the expected behaviour of a mixture of the boiling point of diesel fuel and metathesis product. The results of the measurements can be seen in Fig. 4-4. It can be clearly seen that the influence of the high diesel fuel proportion dominates the boiling curve. The greater the 1-hexene proportion was in the reaction, the lower the boiling curve of the obtained product and the resultant blend. With the mixed fuels, only the distance to the product L resulting from self-metathesis is clearly recognisable, whereby this also lies below the boiling curve of the B20 blend measured as a comparison. All other products lie very close together and hardly deviate from the curve of the pure diesel fuel, particularly in the range up to 60%. The sample metathesis K20 can hardly be distinguished from pure diesel fuel with respect to its boiling curve. The original components of the blend can only be identified by the small steps in the boiling curve above 300 °C.

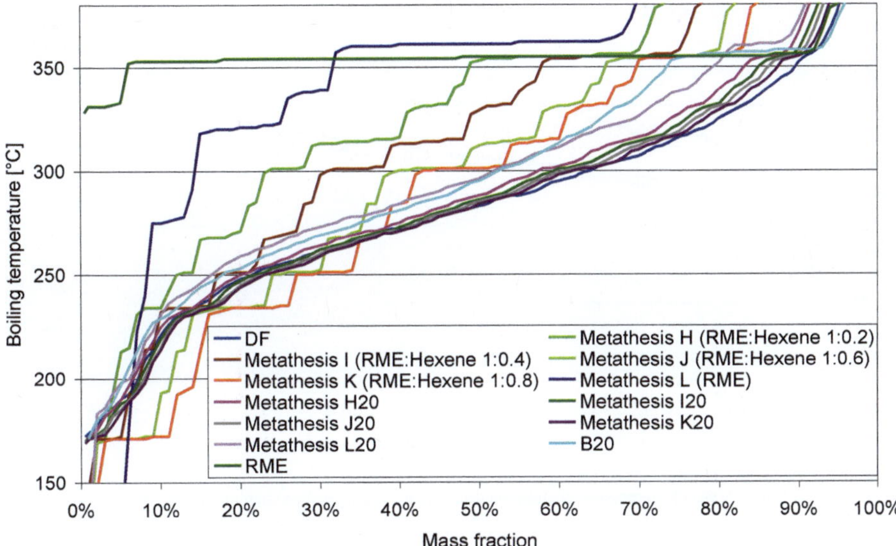

Fig. 4-4:   Comparison of the boiling behaviour of pure metathesis products with their blends

In the procedure that followed, two of the metathesis products were selected for production in larger quantities for operation of a commercial vehicle engine. The selection of these components is explained in greater detail in 4.2.3.

Again, analyses of the boiling curves of these fuels were also made. The corresponding boiling curves can be found in Fig. 4-5. In addition to the metathesis products MM and MN produced in 20-litre quantities, components MK and ML are also reproduced since the new batches were produced in the same way.

In order to show the reproducibility of the production, the boiling curve of the samples should be almost identical. In the illustration it is clear to see that samples L and M and samples K and N correspond with one another and hence reproducibility in production is guaranteed.

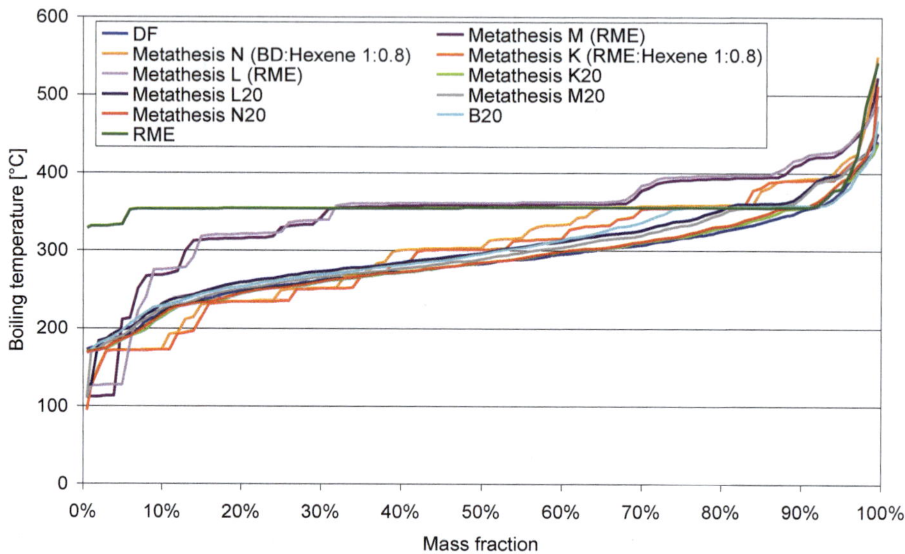

Fig. 4-5:   Boiling curves of the possible metathesis products for use in the commercial vehicle engine

Furthermore, the boiling curves of the blend of 20% metathesis product and 80% diesel fuel are found, since these fuels are used in the engine. In addition, the boiling curve of a B20 blend is presented, which is used for comparison with the mixtures. It is clear that for this blend, the boiling curve of the biodiesel is reached somewhat sooner than is the case with the mixtures with the two metathesis products.

4.1.2 **Miscibility of metathesis fuel with other fuels and engine oil**

In determination of the miscibility, different components have been examined for their behaviour in combination with metathesis products. The matrix for the mixing tests with diesel fuel is shown in Table 4-1. With the mixing of DF and metathesis products, neither turbidity nor the formation of two phases is exhibited. Hence the two components can be mixed without problem. Difficulties first occur when aged fatty acid methyl ester (aging for 40 hours with the introduction of air at 110 °C, Krahl et al., 2009) is added to the mixture. In this case, turbidity of the mixture is only visible as long as the metathesis proportion is 10 to 20%. If the proportion is increased further to 50% or if 5% ethanol or 1-butanol is added, the turbidity can be completely dissipated again.

The behaviour identified here was also exhibited in the past when RME was used instead of the metathesis product (Krahl et al., 2011).

Sample	Ethanol [vol %]	1-Butanol [vol %]	RME aged [vol %]	Ref. DF [vol %]	Metathesis fuel A [vol %]	Mixing behaviour	Sediment formation
1	-	-	-	90	10	miscible	-
2	-	-	-	80	20	miscible	-
3	-	-	-	50	50	miscible	-
4	-	-	-	20	80	miscible	-
5	-	-	10	80	10	miscible	turbidity
6	-	-	10	70	20	miscible	turbidity
7	-	-	10	40	50	miscible	-
8	5	5	10	70	10	miscible	-
9	5	5	10	60	20	miscible	-
10	5	5	10	30	50	miscible	-

Table 4-1:   Reciprocal effects of metathesis products with diesel fuel, RME and alcohols

In a further test series, the diesel fuel proportion in the mixtures was replaced with HVO. The results of the mixing test can be found in Table 4-2. The behaviour described above was also exhibited for the mixtures with HVO.

Sample	Ethanol [vol %]	1- Butanol [vol %]	RME aged [vol %]	HVO [vol %]	Metathesis fuel A [vol %]	Mixing behaviour	Sediment formation
11	-	-	-	90	10	miscible	-
12	-	-	-	80	20	miscible	-
13	-	-	-	50	50	miscible	-
14	-	-	-	20	80	miscible	-
15	-	-	10	80	10	miscible	turbidity
16	-	-	10	70	20	miscible	turbidity
17	-	-	10	40	50	miscible	-
18	5	5	10	70	10	miscible	-
19	5	5	10	60	20	miscible	-
20	5	5	10	30	50	miscible	-

Table 4-2:   Reciprocal effects of metathesis products with HVO, RME and alcohols

Behaviour with engine oil represents the second section with respect to the miscibility test. This investigation is intended to simulate the behaviour of the fuel when entering the engine oil. For this reason, mixtures of 1 mL fuel and 9 mL engine oil in accordance with the temperatures presented in Table 4-3 stored for 24 or 22.5 hours respectively. The samples were then visually inspected, and this was repeated after a further 144 hours at room temperature.

As expected, the viscosity of the mixture changed with temperature: the higher the temperature, the lower the viscosity, which can also be verified by viscosity measurements (SVM 3000 Stabinger viscometer from Anton Paar, TAC of the University of Coburg).

	Metathesis fuel A Cat.: Grubbs 1st generation	Metathesis fuel B Cat.: Umicore M2 60% cross-metathesis, 40% self-metathesis	Metathesis fuel C Cat.: Umicore M2 71% cross-metathesis, 29% self-metathesis
Fuel per sample	1 mL	1 mL	1 mL
Engine oil per sample	9 mL	9 mL	9 mL
Storage (24 h)	-18 °C	-18 °C	-18 °C
Change after temperature effect	increased viscosity	increased viscosity	increased viscosity
Change after a further 144 h at room temperature	visually homogeneous mixture	visually homogeneous mixture	visually homogeneous mixture
Storage (24 h)	RT	RT	RT
Change after temperature effect	none	none	none
Change after a further 144 h at room temperature	visually homogeneous mixture	visually homogeneous mixture	visually homogeneous mixture
Storage (22.5 h)	90 °C	90 °C	90 °C
Change after temperature effect	reduced viscosity; improved intermixing	reduced viscosity; improved intermixing	reduced viscosity; improved intermixing
Change after a further 144 h at room temperature	visually homogeneous mixture	visually homogeneous mixture	visually homogeneous mixture
Storage (22.5 h) open sample vessel	90 °C	90 °C	90 °C
Mass before [g]	7.59	7.68	7.73
Mass after [g]	7.34	7.59	7.67

Table 4-3:   Results of the behaviour of metathesis fuel in engine oil

Due to this viscosity change, improved intermixing of the two involved substances was observed after storage at 90 °C. Turbidity or sediments that were visible to the naked eye did not occur in any of the states. At a temperature of 90 °C the mass of the sample reduces in the course of the test by up to 3% depending on the metathesis product, which is due to evaporation of the volatile components of the metathesis fuel. Similarly with this test, a few fuel components would leave the engine oil again after entry, leading to a reduction in oil dilution and hence possibly to an increase in oil change intervals. Based on this knowledge, problems in the interaction between engine oil and fuel can be excluded over short periods. However, long-term tests were not carried out within the framework of this project.

4.1.3 Material compatibility

In order to obtain a first impression of the behaviour of the new fuel with respect to plastics, a resistance test was carried out on two polymer samples in different fuels.

As already described under 3.1.4, triple determination is the basis of the results. For the polyamide samples (PA) in DF and RME, each one of the three samples led to unrealistic results due to cavities in the material, so these statements are only based on the two other measurements. The mass increase of the samples due to storage in the different fuels is presented in Fig. 4-6. With the reference samples in the climatic chamber at 22 °C and 45% air humidity, the changes remain within the range of standard deviation. In the case of the material polyethylene (PE), there is a clear increase in the sample mass in all three fuels. It is clearly evident that this is almost twice as large for fossil fuel than it is for rapeseed oil methyl ester and metathesis fuel.

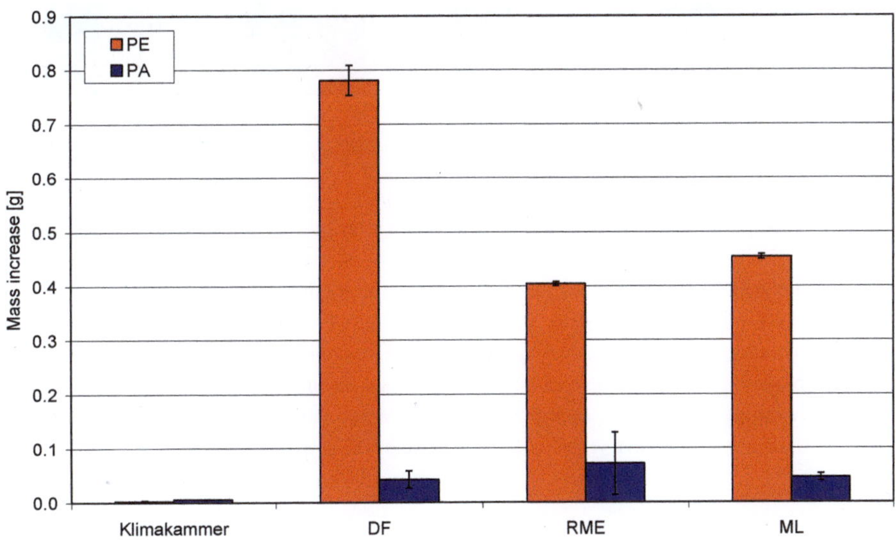

Fig. 4-6:   Mass increase upon storage in fuels with a sample mass of 8.8 g for PE and 11.2 g for PA

Based on the mass of a sample of approx. 8.8 g, the increase is just under 10%. The PA samples expand much less, and with a mass of 11.2 g and fuel storage of less than 0.1 g the result here comprises less than 0.1% of the original mass.

Furthermore, tensile tests were carried out on the samples in the materials technology laboratory of the University of Coburg to determine the modulus of elasticity and tensile strength. The results of these tests can be found in Fig. 4-7 and Fig. 4-8.

It is clear that only small differences are displayed between the uninfluenced samples and the samples that were stored in fuel.

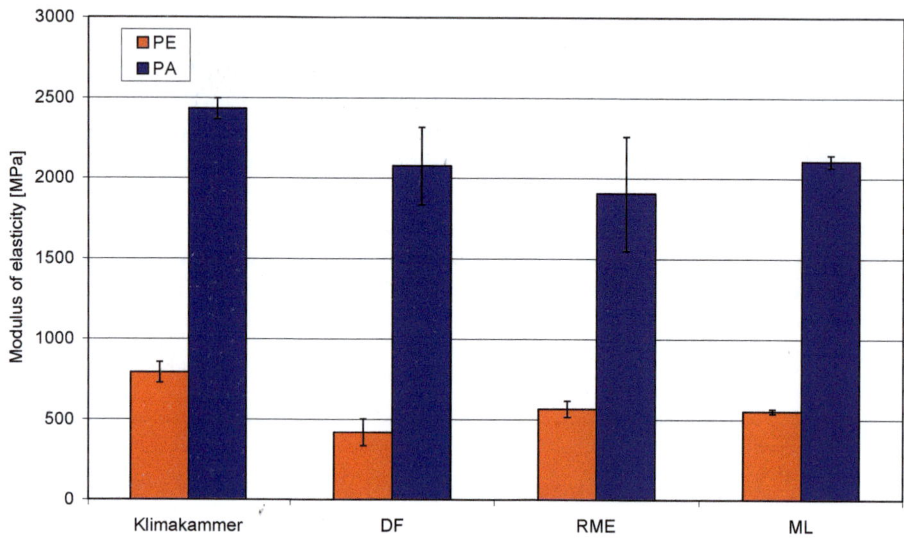

Fig. 4-7:   Influence of storage in fuel on the modulus of elasticity

The modulus of elasticity decreases slightly after storage in fuel. Hence the samples are slightly more pliable than before. This trend is observable for both materials.

There were no significant changes with respect to tensile strength for PA. The reduction in the tensile strength is also minimal when using PE. The results determined here can primarily be based on the sample stored in RME, since according to BAM the resistance of the utilised materials has already been proven. Hence it can be said that the metathesis fuel does not affect the material properties any more strongly than RME. Based on the sample stored in the climatic chamber, changes are caused but these are not decisive with respect to the use of the new fuel in the engine.

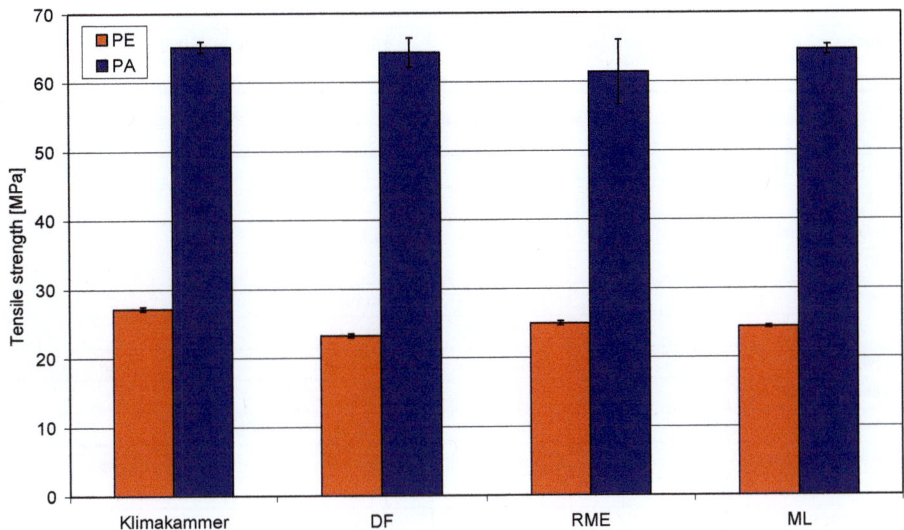

Fig. 4-8:   Influence of storage in fuel on tensile strength

4.1.4 Analysis of standardised fuel properties

For the newly produced metathesis fuel, the question naturally poses itself as to what extent it corresponds with the diesel fuel standard (DIN EN 590). Since a blend of 20% metathesis fuel and 80% reference diesel fuel was used in the engine tests, the analysis was also carried out on this blend. The fuel test was carried out by the company ASG Analytik-Service GmbH. The results are compiled in Table 4-4.

The analysed blend fulfils the majority of the conditions stipulated in the diesel fuel standard. In particular, at 79 °C the flash point is clearly above the required minimum value of 55 °C. Hence the components of the metathesis fuel that boil even at low temperatures do not have a negative influence on the flash point.

The water content is slightly above the guidelines, which could certainly be optimised by changes in fuel production and purification. Also, oxidation stability presents a problem as expected, since the natural and synthetic oxidation stabilisers are decomposed or removed during production of the fuel. This problem could be resolved by the later addition of antioxidants. It remains to be mentioned that with the addition of 20% of the metathesis product to DF, the FAME content understandably ends up above 7% and is 18.1% for the produced blend.

Property	Method	Unit	Threshold values		Test result
	DIN EN		Min.	Max.	
Cetane number	ISO 15195	-	51.0	-	55.7
Cetane index	ISO 4264	-	46.0	-	55.1
Density (15 °C)	ISO 12185	kg/m^3	820	845	838.8
PAH	ISO 12916	% (m/m)	-	8.0	3.5
Sulphur content	ISO 20884	mg/kg	-	10	1.0
Flash point	ISO 2719	°C	> 55	-	79.0
Coke residue	ISO 10370	% (m/m)	-	0.30	0.25
Ash content	ISO 6245	% (m/m)	-	0.01	< 0.005
Water content	ISO 12937	mg/kg	-	200	233
Total contamination	12662	mg/kg	-	24	8
Corrosive effect on copper	ISO 2160	Cor. degree	Class 1		1
Oxidation stability	15751	h	20	-	<0.5
HFRR (at 60 °C)	ISO 12156-1	µm	-	460	266
Kin. viscosity (40 °C)	ISO 3104	mm²/s	2.0	4.5	3.051
CFPP	116	°C	-	0/-10/-20	-17
Course of distillation					
% (V/V) 250 °C	ISO 3405	% (V/V)	-	< 65	20.8
% (V/V) 350 °C	ISO 3405	% (V/V)	85	-	94.3
95% point	ISO 3405	°C	-	360	353.7
FAME content	ISO 14078	% (V/V)	-	7.0	18.1

Table 4-4:   Analysis of metathesis fuel blend MM20 (ASG company)

4.2   Emissions testing on the single cylinder test engine

For evaluation of the fuels, determination of their emissions is indispensable. Since the different metathesis components could only be produced on a scale of one litre, the first emissions measurements took place on the single-cylinder test engine. With these tests, the focus was on the regulated exhaust gas components. Non-regulated exhaust gas components such as PAH, mutagenicity and carbonyls were only determined with selected fuels.

4.2.1   Regulated emissions

Regulated emissions include both the gaseous exhaust gas components NO_x, HC and CO and also the particles contained in the exhaust gas. The nitrogen oxide emissions of the utilised fuels are presented in Fig. 4-9. Fossil diesel fuel and rapeseed oil methyl ester serve as comparison fuels. Since the metathesis fuels were used as a blend with 20% metathesis proportion, the list of comparison fuels expanded by a B20 blend of DF and RME.

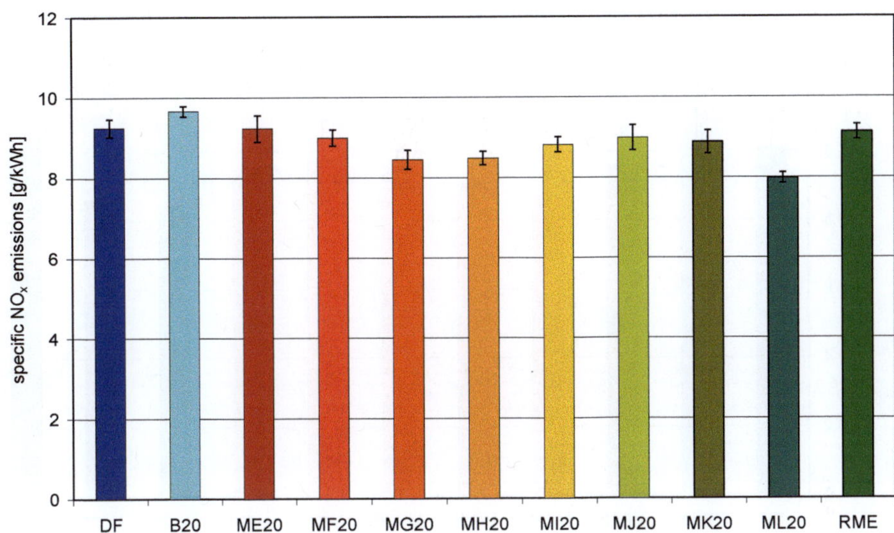

Fig. 4-9:   Specific NO$_x$ emissions of the metathesis fuels in the Farymann single-cylinder engine in the five-point test

As already known from earlier measurements (Schröder et al., 1998), the use of RME with 9.12 g/kWh in the Farymann engine does not lead to higher nitrogen oxide emissions than DF or B20. The differences between DF and RME are not significant. On the other hand, the B20 blend leads to a clear increase in nitrogen oxide emissions. All metathesis blends show lower NO$_x$ values than the comparison fuels. Metathesis product ML exhibits the lowest nitrogen oxide emissions with 7.98 g/kWh. However, due to standard deviations no significant differences can be determined between a majority of the metathesis fuels.

The measurement results for the particle mass in the exhaust gas can be found in Fig. 4-10. The differences between the individual comparison fuels are very low and lie within the range of the standard deviation. Since these are much higher than e.g. with NO$_x$ or CO with the particle mass determination, in this case no significant differences can be determined between the individual metathesis components. MK delivers the lowest value for particle mass with 0.84 g/kWh and hence clearly lies below the three other comparison fuels with 1.01 g/kWh for DF, 1.04 g/kWh for RME and 1.05 g/kWh for B20.

The hydrocarbon emissions follow in Fig. 4-11. Here, RME has by far the lowest figure with 1.34 g/kWh, followed by B20 with 1.57 g/kWh. All metathesis fuels and also DF were considerably higher. However, the differences between DF and the metathesis products are not significant due to the very large standard deviations. Based on these results, it can be assumed that biodiesel's advantage of lower hydrocarbon emissions is not transferred to the metathesis products. This must naturally be confirmed in tests on the commercial vehicle engine.

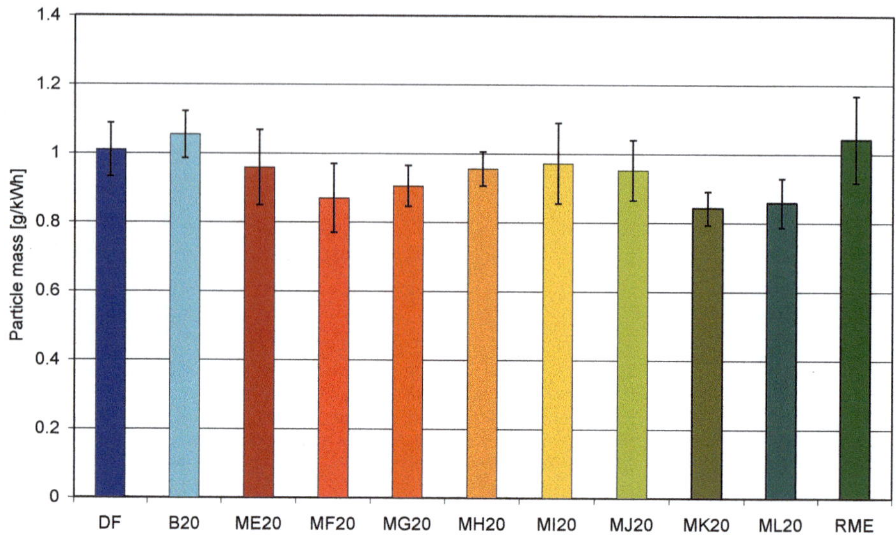

Fig. 4-10:   Particle mass emissions of metathesis fuels in the Farymann single-cylinder engine in the five-point test

Furthermore it can be seen that with the HC emissions with ME and MF, other metathesis components exhibit advantages than with the nitrogen oxides and particle mass.

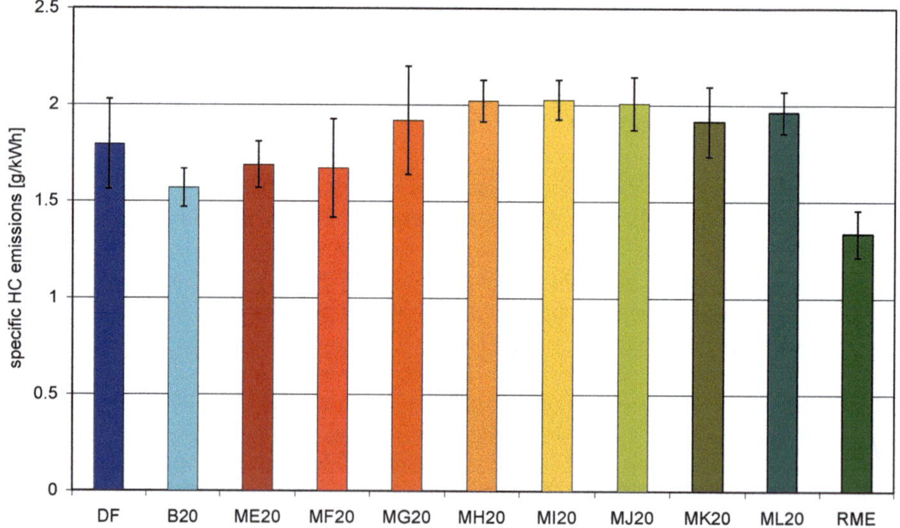

Fig. 4-11:   Specific HC emissions of the metathesis fuels in the Farymann single-cylinder engine in the five-point test

With the carbon monoxide emissions, the fourth of the regulated exhaust gas components is presented in Fig. 4-12. A similar picture is shown here than with the hydrocarbons. The lowest emissions are found with RME and B20. With 7.21 g/kWh, DF is in the same magnitude as most metathesis fuels, wherein ML20 is found much lower than the others with 6.33 g/kWh and lies in the area of RME and B20.

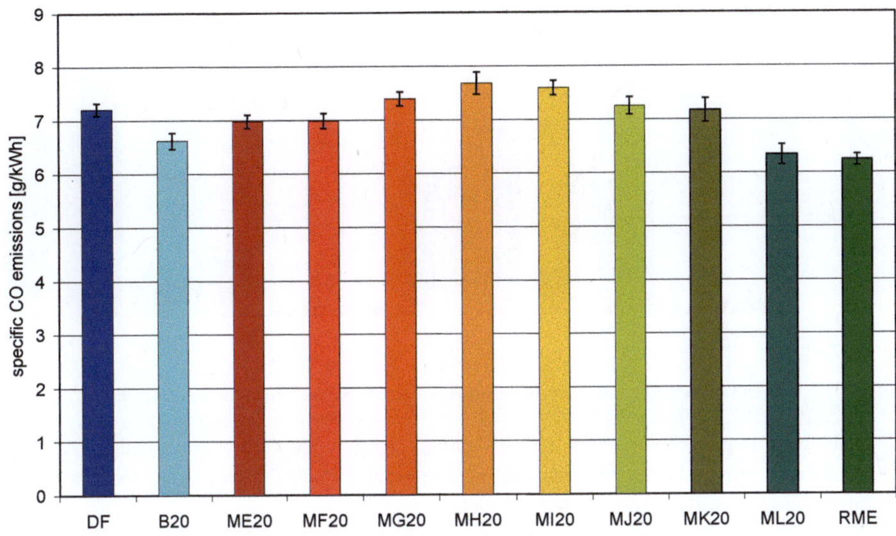

Fig. 4-12:   Specific CO emissions of metathesis fuels in the Farymann single-cylinder engine in the five-point test

Since the standard deviations are much lower with the CO measurement, in part significant differences can be found here between the different metathesis products. Hence ML20 leads to significantly lower CO emissions than MG20, MH20 and MI20.

4.2.2 Non-regulated emissions

With regard to the comparison and selection of metathesis products for the further project steps on the test engine, the non-regulated emissions are of secondary importance since they are not intended to feature in the selection process of the fuels for the tests on the commercial vehicle engine. Hence only a few selected fuels were tested for carbonyls and mutagenicity. Only the determination of PAH emissions took place for all utilised metathesis components.

For this reason, it was decided to commence with observation of the PAH emissions of particulates and condensates and the resultant comparison of all tested fuels. The PAH masses contained in the particulates of the individual samples can be found in Fig. 4-13 and Fig. 4-14. Hardly any differences can be discerned between the individual fuels.

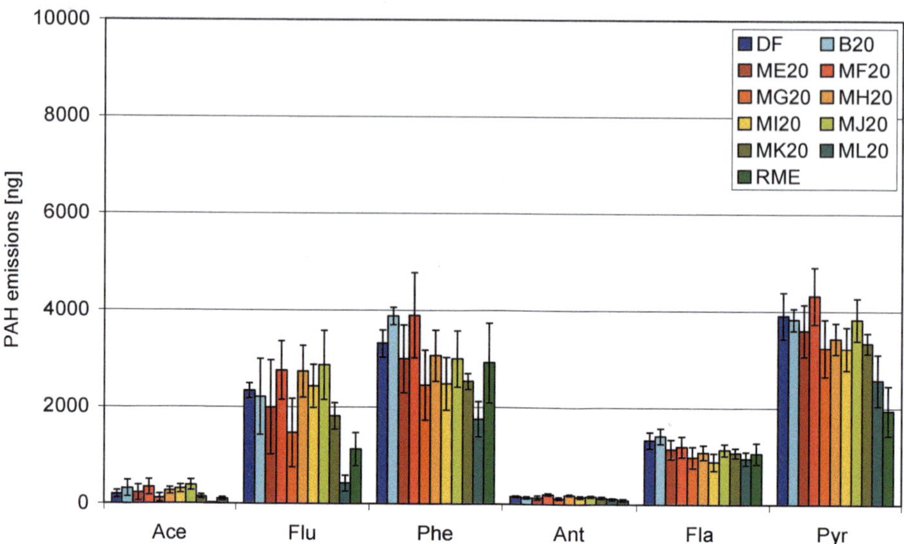

Fig. 4-13:   PAH emissions in the particulates in the Farymann single-cylinder engine in the five-point test

DF tends towards greater emissions than RME, especially in the case of the smaller PAH (Fig. 4-13). However, for the larger PAH that are more relevant from health viewpoints there are hardly any differences between the two comparison fuels.

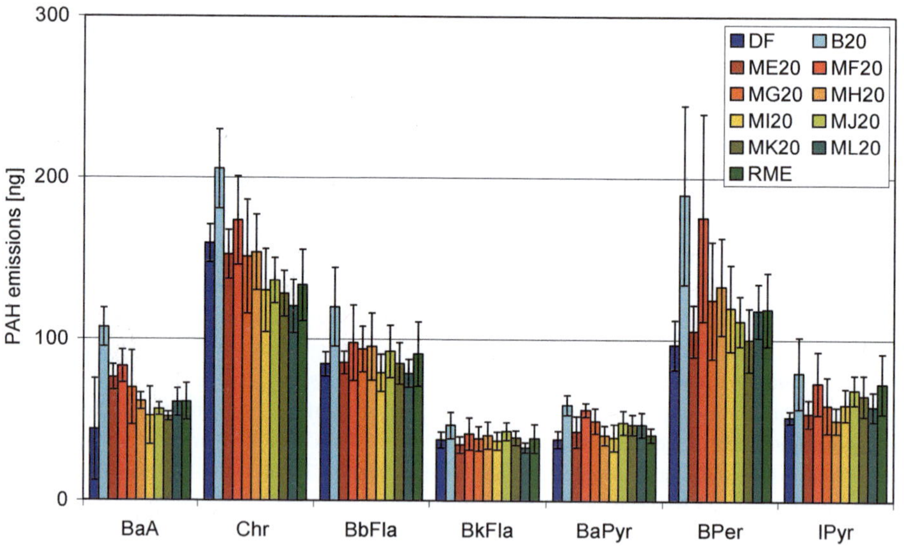

Fig. 4-14:   PAH emissions in the particulates in the Farymann single-cylinder engine in the five-point test

The variances between the fossil diesel fuel and the B20 blend and also the metathesis fuel blends are too small to allow statements to be made on the advantages and disadvantages of the individual fuels. Only in a few cases of the results presented in Fig. 4-14 do the values for B20 appear slightly increased. However, precisely these measurements also have high standard deviations and hence also large variations in the result, which clearly reduces the validity and hence also does not allow a clear conclusion.

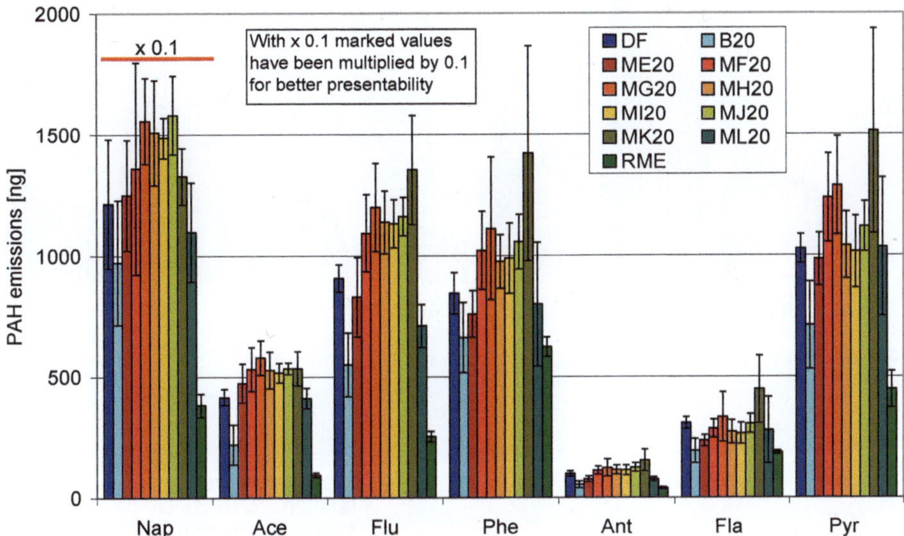

Fig. 4-15:   PAH emissions in the condensates in the Farymann single-cylinder engine in the five-point test

The measurements of the condensates presented in Fig. 4-15 and Fig. 4-16 clearly show that the PAH masses contained in them are much lower than those in the particulate.

In the particulate sample, B20 tended to show slightly-increased PAH concentrations. With the condensate samples, the tendency is the exact opposite.

Addition of the emissions from particulate and condensate results in Fig. 4-17 and Fig. 4-18. Here, it is clear that the emissions for RME and ML20 tend to be lower for some PAH such as fluorene and pyrene. In contrast, B20 and MF20 frequently belong to the fuels with the highest emissions.

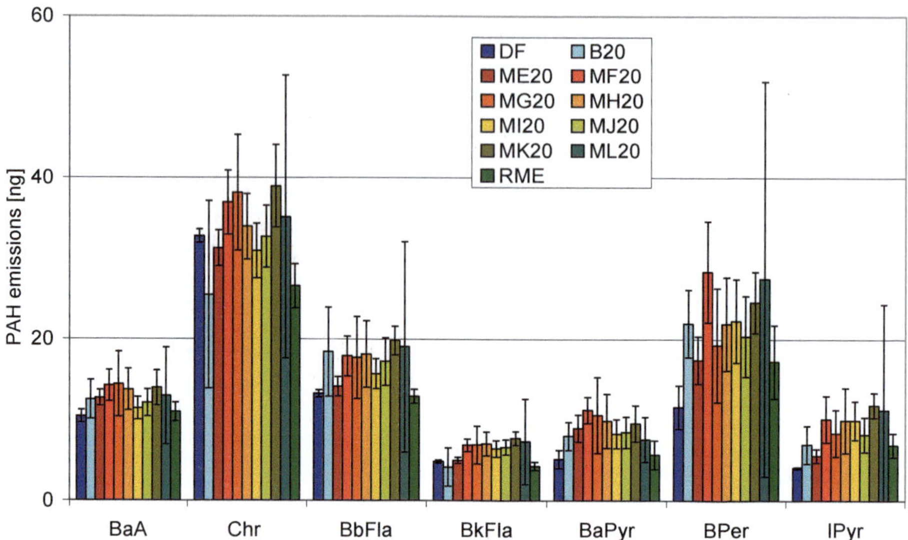

Fig. 4-16: PAH emissions in the condensates in the Farymann single-cylinder engine in the five-point test

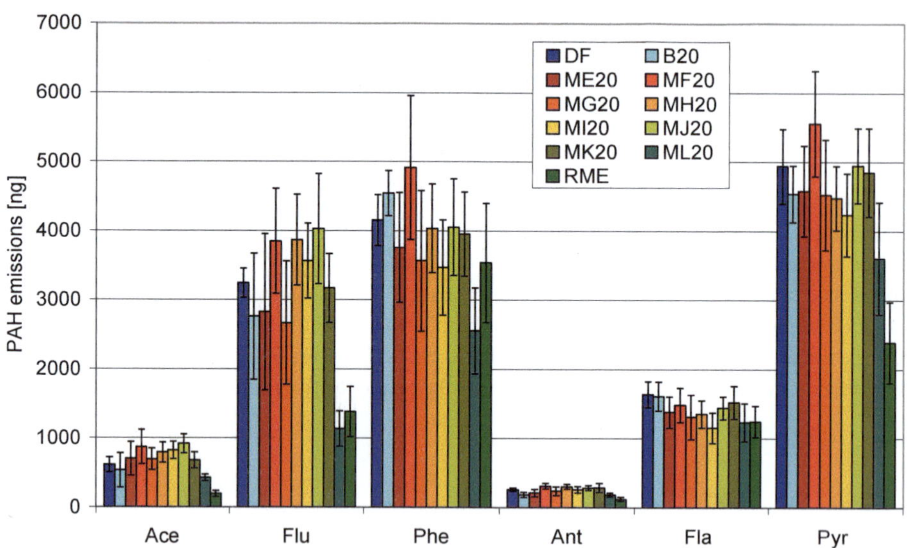

Fig. 4-17: Total of PAH emissions from particulate and condensate in the Farymann single-cylinder engine in the five-point test

The remaining fuels exhibit only minimal differences from one another that are within the magnitude of standard deviations.

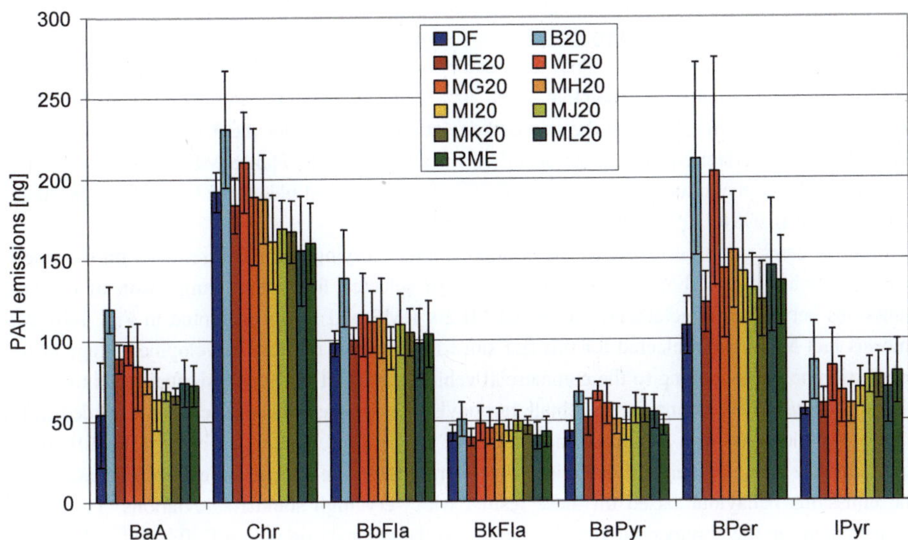

Fig. 4-18:   Total of PAH emissions from particulate and condensate in the Farymann single-cylinder engine in the five-point test

There are also no great differences between the 20% blend and the diesel fuel, whereby the DF provides somewhat lower values for the PAH that are relevant to the effect (Fig. 4-18).

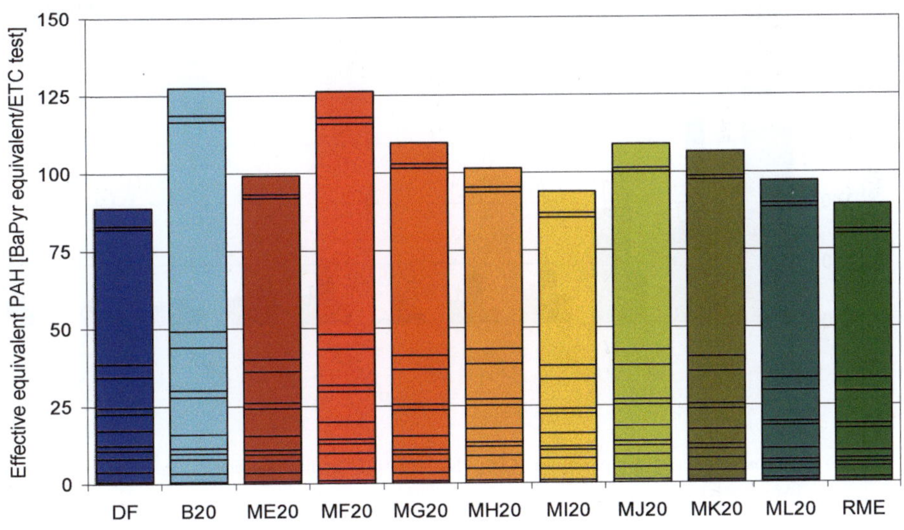

Fig. 4-19:   Effect equivalent of PAH emissions from particulate and condensate in the Farymann single-cylinder engine in the five-point test

Observation of the effect of the PAH emissions took place by adding the presented emissions as effect equivalent of the benzo[a]pyrene. The sums for the individual fuels are plotted in Fig. 4-19. The differences between the individual fuels can be seen more clearly here. The effective potential of RME and DF is slightly less than that of the 20% blend. In total, B20 and MF20 clearly lie above the remaining fuels. In relation to the standard deviations available for the PAH measurement, only tendencies in the effect equivalent are specified for all other fuels. Hence ME20, MI20 and ML20 lie in the order of magnitude of DF and RME. The metathesis fuel blends MG20, MH20, MJ20 and MK20 lie with 10% to 20% higher equivalents than DF.

In order to carry out an estimate of the emissions after carrying out measurements on the single-cylinder engine, the carbonyls were determined for selected fuels. The comparison of carbonyl emissions between the metathesis fuel blend ML20, RME and DF is presented in Fig. 4-20. Metathesis fuel ML20 was selected for this test since it represents the final development stage in fuel production before switching to the commercial vehicle engine. The metathesis fuel tends to exhibit slightly increased emissions for the small carbonyls. The greatest increase can be recorded in the case of acetaldehyde with an approximate factor of 2. For the remaining carbonyls, the differences are within the range of standard deviation. However, a definitive conclusion cannot be drawn about the emissions behaviour based on these results with very high standard deviations. To do this, measurement on the commercial vehicle engine and the comparison with a B20 blend of RME and DF that corresponds with the metathesis blend is necessary.

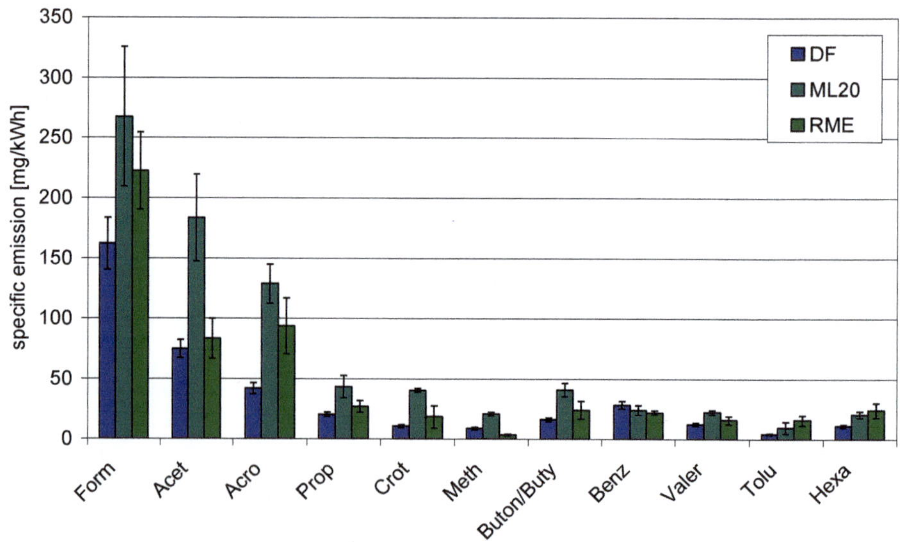

Fig. 4-20: Comparison of carbonyl emissions of DF, RME and ML20 in the Farymann single-cylinder engine in the five-point test

A further, non-regulated evaluation criterion for the engine exhaust gas is mutagenicity. Based on the presented mutagenicity results, it can be estimated as to whether this effect of the emissions could already be an exclusion criterion for the altered fuels. The analysis was carried out both in condensate and in particulate. In addition, two different bacterial strains were used. The number of mutations per sample with the TA98 strain is shown in Fig. 4-21. In part, clear differences between the individual fuels can be seen. The lowest mutagenic potential is borne by the B20 blend of DF and RME. Therefore, for the Farymann single-cylinder engine used here, there is no maximum for the mutagenicity with B20, as was the case in earlier measurements in other engines (Krahl et al., 2008).

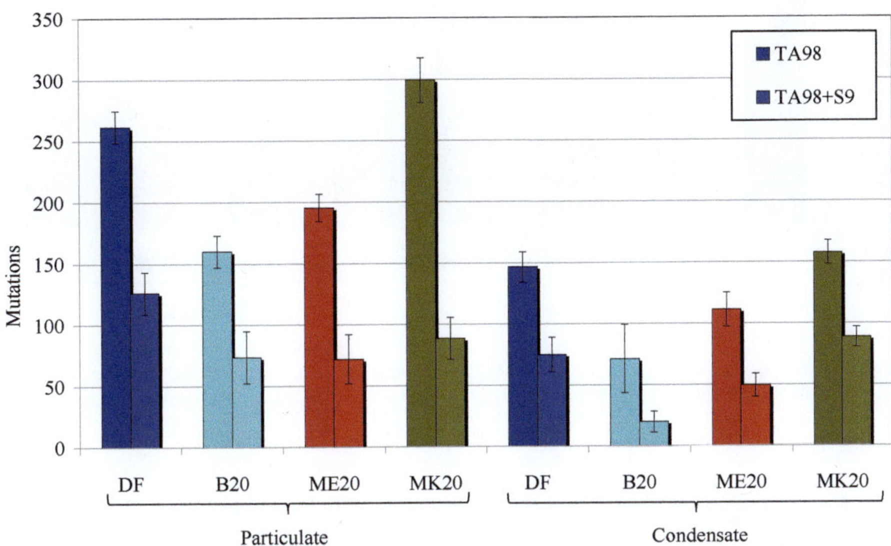

Fig. 4-21:   Mutagenicity of DF, B20, ME20 and MK20 on combustion in the Farymann single-cylinder engine (TA 98 +/- S9)

This could be due to further developments in the area of the fuel filter and the resultant precipitation of oligomers. In contrast, the fuel ME20 that is produced by self-metathesis leads to a slightly-increased number of mutations. In part, the product MK20 that is produced by cross-metathesis with 1-hexene shows a significant increase in mutations both for the particulate and also the condensate and hence lies slightly above the diesel fuel. The TA100 strain tends towards the same behaviour (cf. Fig. 4-22). Here too, B20 and ME20 lie below the other two fuels. However, contrary to the first graphic there are no great differences between B20 and ME20 and between DF and MK20.

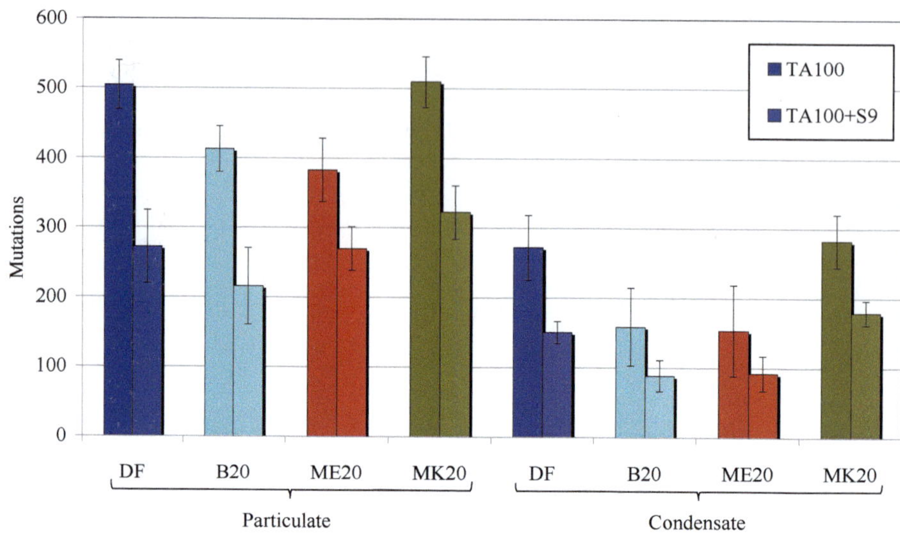

Fig. 4-22:   Mutagenicity of DF, B20, ME20 and MK20 on combustion in the Farymann single-cylinder engine (TA 98 +/- S9)

4.2.3 Fuel selection for the tests in the commercial vehicle engine

A much greater fuel volume of approx. 20 litres was required for continuation of the measurements in the commercial vehicle engine. Since such quantities could not be produced for 14 different fuels, the two best-suited fuels were to be selected from the tests on the single-cylinder engine. Different criteria were used to make this decision. For one, evaluation of the regulated emissions was carried out. In doing this, the nitrogen oxides and the particle mass were more strongly weighted since adhering to of the limits represents a particular challenge for engine manufacturers, especially for these two components. Other indications for the fuel selection were the ratio of biodiesel to 1-hexene and the boiling curve of the metathesis product.

The evaluation of the individual fuels displayed in Table 4-5 results from these criteria. The first four products MA to MD are not taken into consideration. No complete engine tests were carried out with these first fuels since the produced volumes were still very low and the purification steps have only just been developed and tested. Component ML is also not listed since it was created at a later point in time as a further development of product ME.

In the area of emissions, the sequence was produced on the basis of nominal values. Here, no evaluation takes place with regard to significance. The respective lowest value of a component is evaluated as 1 and the highest value as 7. With the biodiesel proportion, the maximum biodiesel content in the metathesis reaction receives 1, and with the boiling curve 1 indicates the boiling behaviour that is closest to that of diesel fuel. The weighting factors for the individual criteria can be found in the first row.

	Biodiesel proportion	Boiling curve	NO$_x$	PM	CO	HC	Total
Sample number	0.15	0.15	0.25	0.25	0.1	0.1	1
MA	-	-	-	-	-	-	-
MB	-	-	-	-	-	-	-
MC	-	-	-	-	-	-	-
MD	-	-	-	-	-	-	-
ME	1	7	7	6	1	2	4.75
MF	7	1	6	2	2	1	3.50
MG	4	4	1	3	5	4	3.10
MH	2	6	2	5	7	7	4.35
MI	3	5	3	7	6	6	4.90
MJ	5	3	5	4	4	5	4.35
MK	6	2	4	1	3	3	3.05

Table 4-5:   Evaluation of the metathesis products according to emissions, biodiesel proportion and boiling curve

Due to this ranking and the specified weighting, metathesis product MK showed the best properties and was therefore selected for further use in the project. Since the biodiesel content in this fuel is rather low with 1 eq biodiesel to 0.8 eq hexene, a second fuel should be selected where the weighting is clearly on the side of the biodiesel proportion. With these changed conditions, the decision for the second fuel is presented completely differently. The new selection conditions are listed in Table 4-6. Hence metathesis product ME was selected as a second fuel for further measurement.

	Biodiesel proportion	Boiling curve	NO$_x$	PM	CO	HC	Total
Sample number	0.5	0.1	0.1	0.1	0.1	0.1	1
ME	1	7	7	6	1	2	2.80
MF	7	1	6	2	2	1	4.70
MG	4	4	1	3	5	4	3.70
MH	2	6	2	5	7	7	3.70
ML	3	5	3	7	6	6	4.20
MJ	5	3	5	4	4	5	4.60
MK	6	2	4	1	3	3	4.30

Table 4-6:   Evaluation of the metathesis products with 50% weighting of the biodiesel proportion

Here, this is a product that was completely produced via self-metathesis and hence has a biodiesel proportion of 100%. However, it is also clear that this fuel exhibits great differences to diesel fuel with respect to the boiling curve (Fig. 4-3).

4.3   Emissions testing in the commercial vehicle engine

In order to be able to correctly evaluate a fuel, emissions tests on state-of-the-art engines are also required. For this reason, the two metathesis fuels selected under 4.2.3 were produced in 20-litre quantities and tested for their emissions as 20% blends in the OM 904 LA commercial vehicle engine. In addition to the regulated emissions nitrogen oxides, particulate matter, carbon monoxide and hydrocarbons, non-regulated components of the exhaust gas - namely ammonia, particle size distribution, PAH and carbonyls were determined and the mutagenic effect of the exhaust gas was analysed. In addition to the two metathesis blends, measurements were again carried out with reference diesel fuel, rapeseed oil methyl ester and a B20 blend of these two fuels.

4.3.1 Regulated emissions

As already undertaken with the single-cylinder test engine, the regulated emissions were also determined in the OM 904 LA commercial vehicle engine. The ETC test was used for the measurements. On observation of the nitrogen oxide emissions, a look at AdBlue dosage also took place with this engine due to the dependence of the SCR catalytic converter on the mass flow of the reducing agent.

As to be expected, the nitrogen oxide measurement in Fig. 4-23 exhibited substantially higher emissions known from a multitude of tests (Lapuerta et al., 2008) during engine operation with RME.

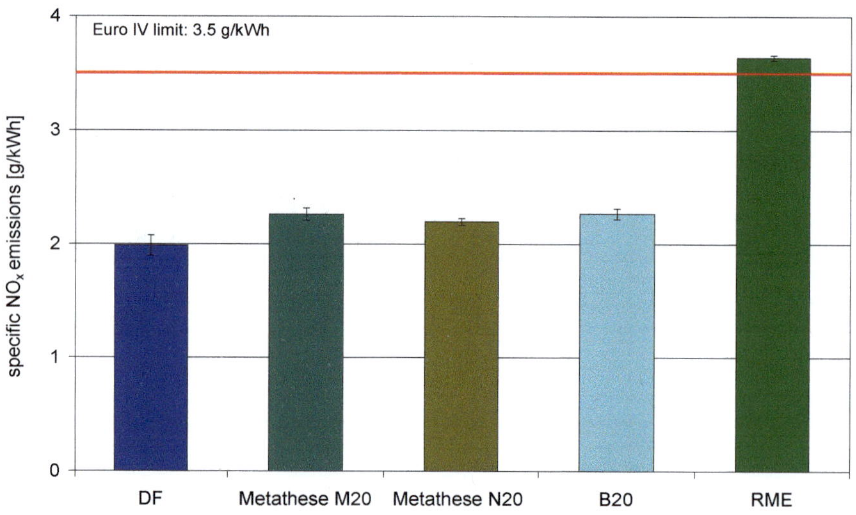

Fig. 4-23:   Specific NO$_x$ emissions of the OM 904 LA in the ETC test

The standard for Euro IV engines of 3.5 g/kWh was not fulfilled with the measured 3.64 g/kWh. The differences between DF and the three fuel blends are much smaller. The value of the diesel fuel is significantly lower than the blends at 1.99 g/kWh. No significant differences can be detected between the B20 blend and the metathesis fuels. Therefore in the range of nitrogen oxides, no influence can be identified of the boiling point on emissions.

If the associated AdBlue mass flows in Fig. 4-24 are considered, only very small deviations can be detected. The AdBlue dosage is set by the engine control unit depending on the load point and the catalyst temperature.

Hence dosage is independent of the fuel as long as the fuel only has a minimal effect on the exhaust gas temperature. A change in dose takes place with the catalyst temperatures in 4 °C increments.

The difference between the maximum dosage for MN20 of 484 g/h and the minimum dosage for RME of 470 g/h is just 14 g/h. At 6 g/h, the standard deviation of all measured values is 1.2% of the absolute value and hence hardly above the deviations between the tests with the same fuel.

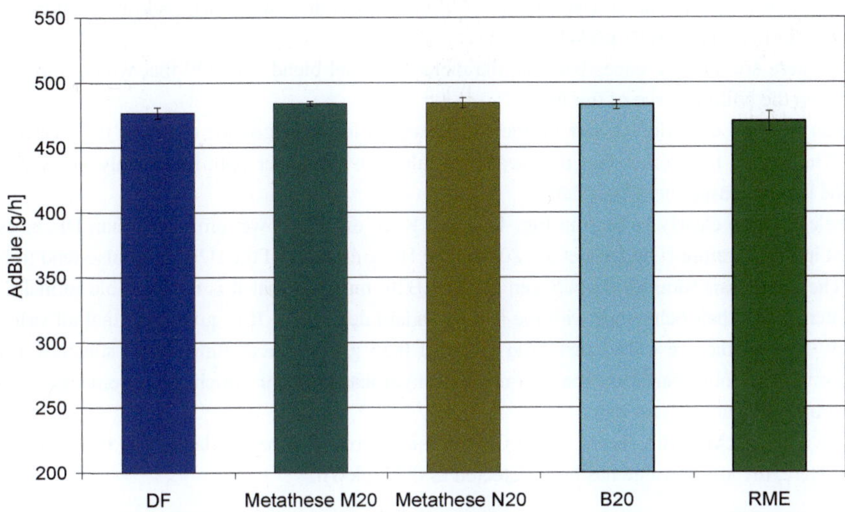

Fig. 4-24:   AdBlue dosage of the OM 904 LA in the ETC test

Hence it can be assumed that the differences in the dosage are too small to have a decisive effect on nitrogen oxide emissions.

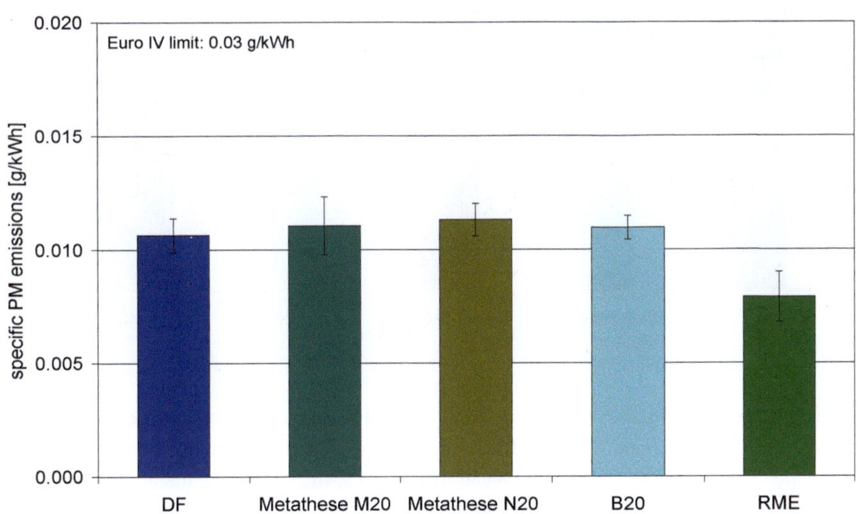

Fig. 4-25:   Specific PM emissions of the OM 904 LA in the ETC test

The behaviour of low PM emissions for RME that is already established in the literature (Lapuerta et al., 2008) is found with the analysis of particulate matter in Fig. 4-25.

However, only minimal differences are found between the remaining four fuels, so it is possible to speak of practically the same emissions here. All measured values lie clearly below the threshold value valid for the ETC of 0.03 g/kWh.

Here too, there are no differences between the metathesis fuel blend and B20 that would imply an influence on the boiling curve of the fuel.

Very large standard deviations occur for the HC emissions in the magnitude of 8% of the measured values. This is due to the fact that the measured values in the lower ppm range only lie slightly above the lower measurement boundary.

Nevertheless, it can clearly be seen in Fig. 4-26 that RME exhibits lower emissions than DF, as established in the literature (Lapuerta et al., 2008). The HC emissions of the B20 blend also tend to be lower. The metathesis samples lie between DF and B20, meaning that it is not possible to make a clear statement on their behaviour with the large standard deviations. It is quite clear that all values lie below 10% of the prescribed threshold value of 0.55 g/kWh. According to the standard, this boundary value is only based on the non-methane hydrocarbons, for which measurements of the total hydrocarbons are present here.

The behaviour of RME with regard to the CO emissions proves to be similar to that of the hydrocarbons. Here, the lowest value has been detected as 0.77 g/kWh.

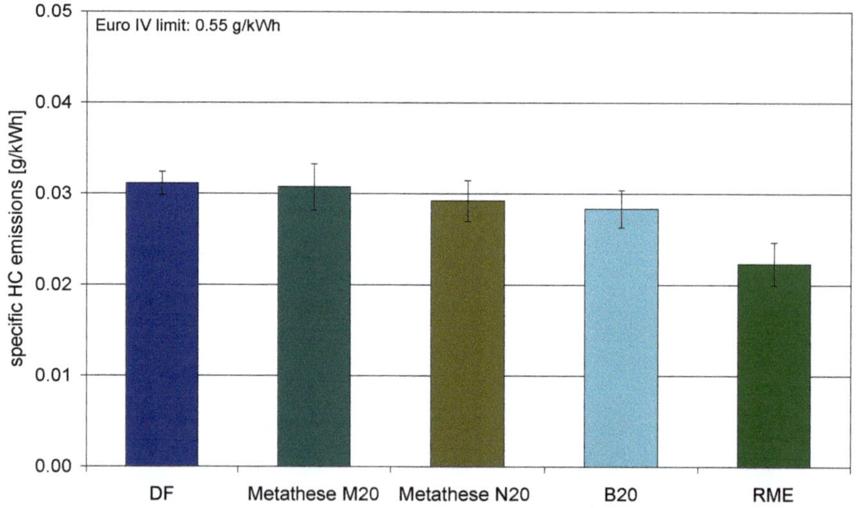

Fig. 4-26:   Specific HC emissions of the OM 904 LA in the ETC test

However, it is clear from Fig. 4-27 that the B20 and MN20 lead to an increase in measured values based on DF. On the other hand, the MM20 blend delivers results in the magnitude of the diesel fuel. If reference is made to the comparable blend of DF and RME, one of the metathesis fuels now exhibits a reduction in carbon monoxide emissions.

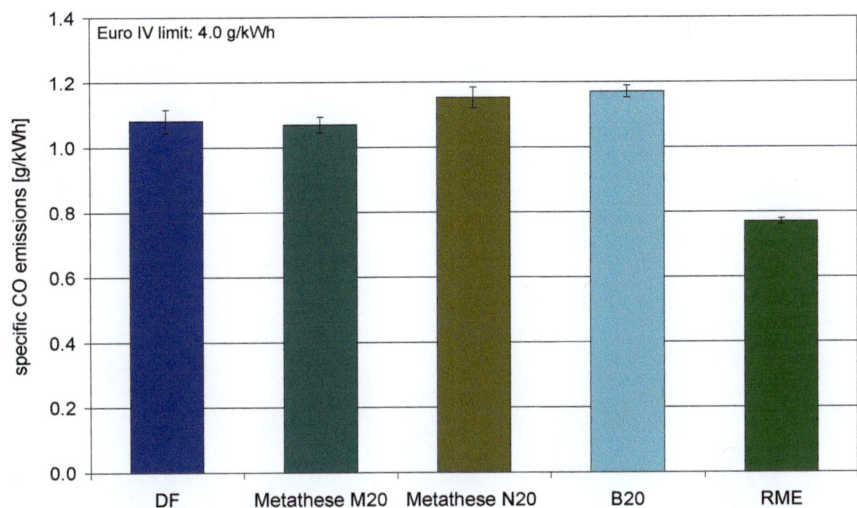

Fig. 4-27:   Specific CO emissions of the OM 904 LA in the ETC test

Since the energy content of the utilised fuels is not the same, the average output of the engine in the ETC test and the fuel consumption are shown in the following illustrations.

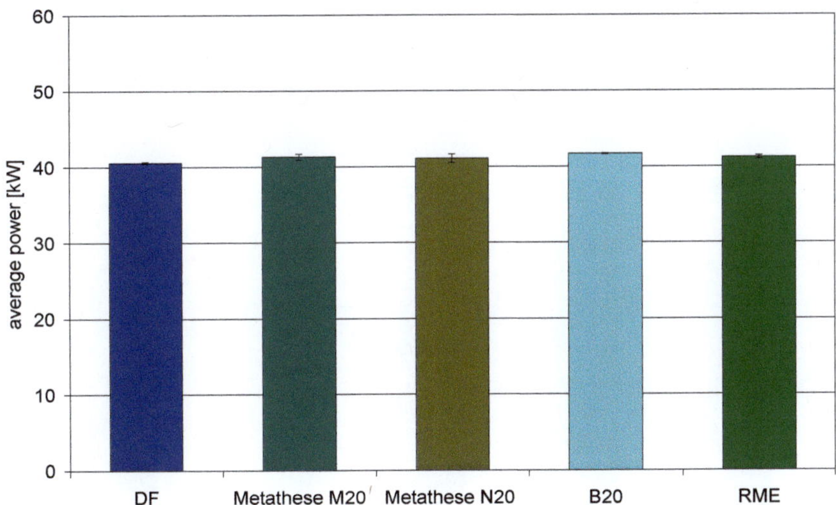

Fig. 4-28:   Average output of the OM 904 LA in the ETC test

It is clear in Fig. 4-28 that the engine delivers almost constant output over the test cycle, independent of the fuel, which is also understandable due to the very low full-load proportions in the ETC test. No dependency on the heating value that could restrict the maximum output at maximum injection quantity can be discerned.

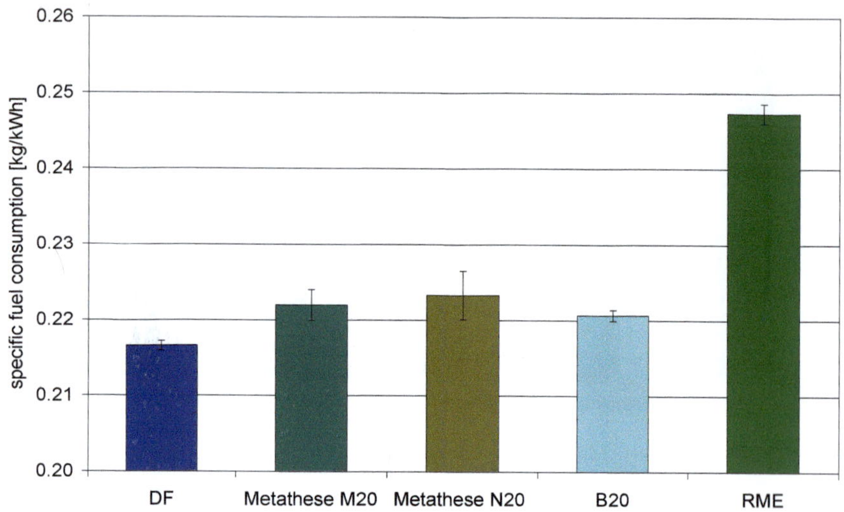

Fig. 4-29:   Specific fuel consumption of the OM 904 LA in the ETC test

In contrast to this, it becomes clear in Fig. 4-29 that the consumed masses vary depending on fuel.

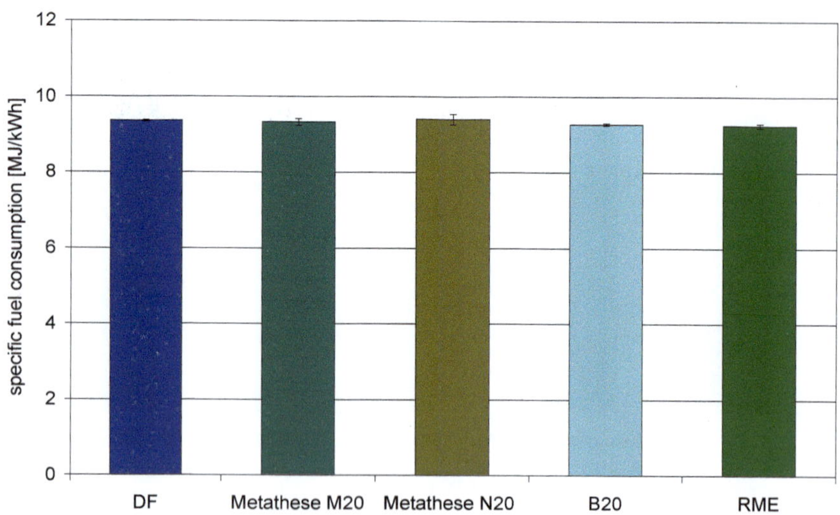

Fig. 4-30:   Heating value-adjusted fuel consumption of the OM 904 LA in the ETC test

On the one hand, the greater density of the RME has an effect here that leads to increased mass with the same volume of fuel. On the other hand, the heating values of the fuels play a role here. In this way, attainment of the output for RME specified in the test cycle requires the injection of an increased volume of fuel since the energy content of the fuel is lower.

Since the blends contain a 20% proportion of biofuel with lower energy content, an increase in the required mass can also be ascribed to them. If one adjusts the results by including the respective heating values of the fuel, then significant differences in consumption can no longer be detected (Fig. 4-30). For the correction, the heating values for the metathesis fuels and the RME were calculated from the C-H-O ratios and then the values were determined for the blends in correspondence with the percentile composition.

4.3.2 Non-regulated emissions

In addition to the regulated emissions, an analysis of many non-regulated exhaust gas components was also carried out. In the process, tests were carried out on particle size distribution, ammonia emissions, polycyclic aromatic hydrocarbons and the mutagenicity of the exhaust gas. Firstly, the results of the particle size distribution and the NH_3 emissions that will be incorporated into legislation from Euro VI are to be taken into consideration.

The particle size distribution of the tested five fuels determined by ELPI is represented in Fig. 4-31. As to be expected, due to the much lower particulate matter determined regarding the regulated emissions (Fig. 4-25), lower particle emissions are exhibited with RME than with the other fuels. This difference can be observed over the entire size range. Even the 20% biofuel in the blends leads to a clear reduction in particle count. The differences between the individual blends are clearly lower.

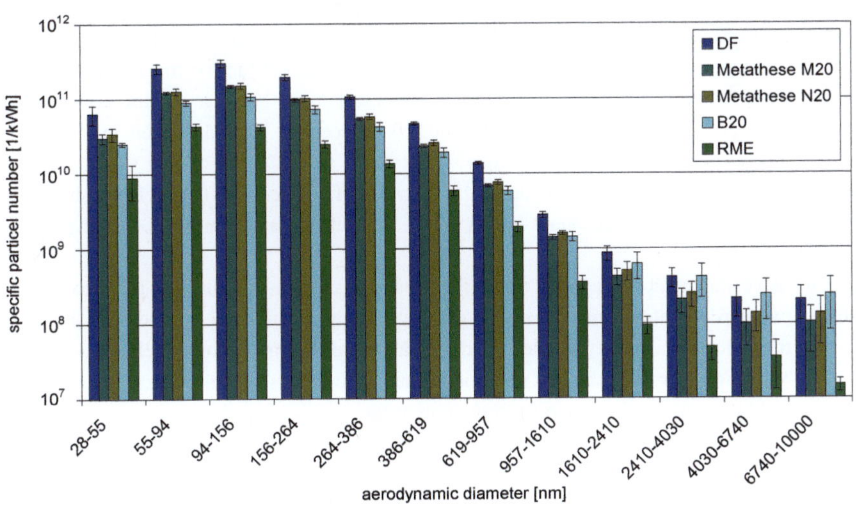

Fig. 4-31:   Particle size distribution of the OM 904 LA in the ETC test, measured via ELPI

The blends of metathesis fuels appear to produce somewhat higher particle counts than the comparable blend with RME. The deviations between the individual measurements clearly increase, especially in the range above 1 μm, leading to an increase in standard deviations. This is due to the very low particle concentrations in this measurement range.

Observation of the ammonia emissions follows in the area of non-regulated exhaust gas components, which have played a role since introduction of SCR technology. A boundary value of 10 ppm is specified from the Euro VI standard.

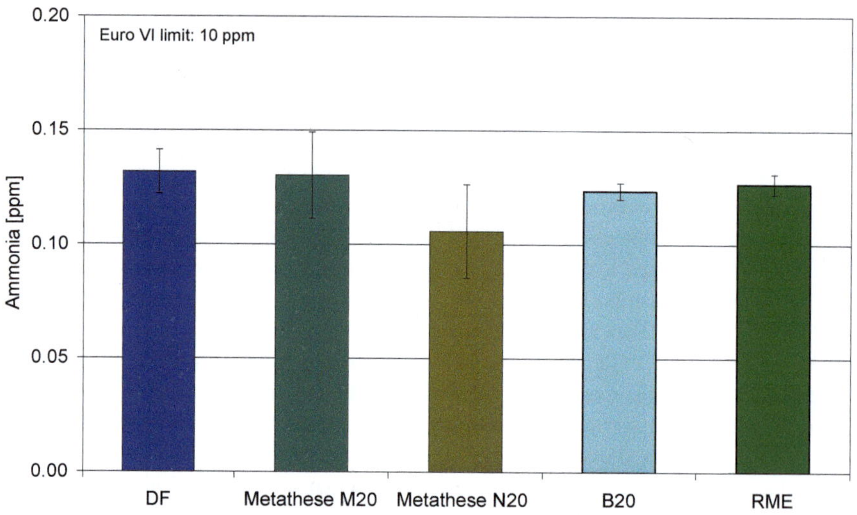

Fig. 4-32:   Ammonia emissions of the OM 904 LA in the ETC test

The emitted ammonia quantities during the test series are shown in Fig. 4-32. For all five fuels, emissions lie between 0.1 and 0.15 ppm and are hence very clearly above the future boundary value. Since the detected values lie at the lower measurement range boundary of the utilised mass spectrometer, no significant differences can be identified between the individual fuels in spite of the low standard deviation in the order of magnitude of 20 ppb.

A statement about the ratio of ammonia slip of the individual fuels between themselves can only be made with increased dosage and resultant higher emissions, but this is not intended with the utilised engine.

Hence the decisive statement is that none of the fuels leads to significant emissions of ammonia and therefore all five can be categorised as completely non-critical in this area. In addition, the result clarifies that the reduction capacity of the dosed AdBlue mass is almost completely consumed by the SCR catalytic converter and a further reduction in nitrogen oxides at the catalytic converter can only be achieved by increased dosage of the reduction agent.

The results of the PAH analysis are presented in the following illustrations. The larger PAH are more prevalent in the particulates (Fig. 4-33), whereas the smaller ones are more prevalent in the condensate (Fig. 4-34). The PAHs that are not listed in the illustrations are present in such small concentrations that it is no longer possible to prove them beyond doubt within the scope of the standard deviations .

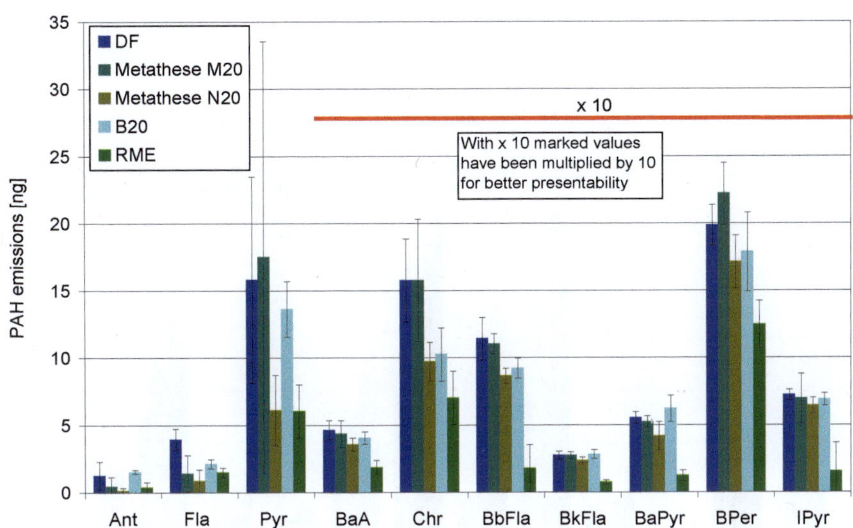

Fig. 4-33:   PAH emissions in the particulate of the OM 904 LA in the ETC test

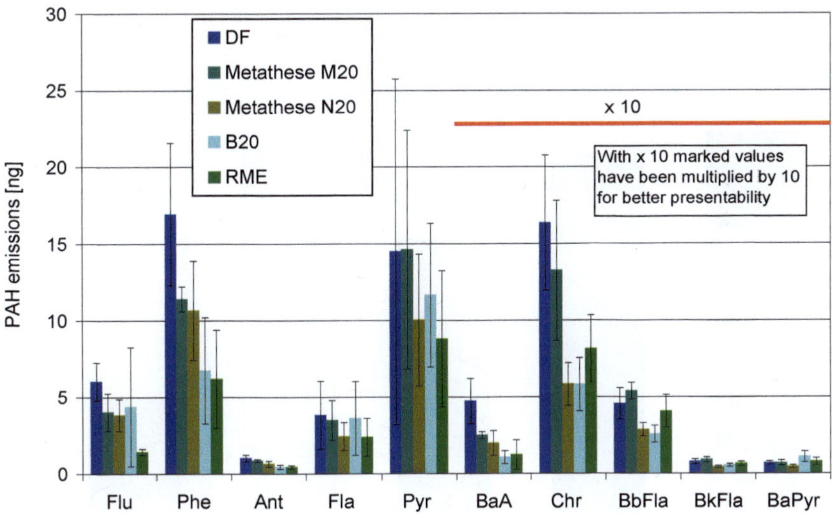

Fig. 4-34:   PAH emissions in the condensates of the OM 904 LA in the ETC test

Only small differences are exhibited between the individual tested blend fuels, but these always lie within the magnitude of the standard deviation. Only RME as a comparison fuel tends towards lower PAH emissions, as can be seen in the totals display for particulate and condensate in Fig. 4-35.

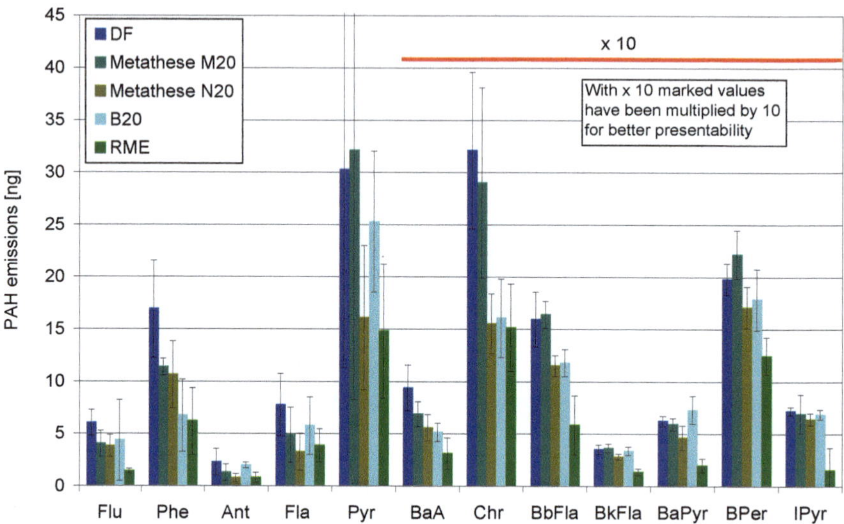

Fig. 4-35: Total of PAH emissions for particulate and condensate of the OM 904 LA in the ETC test

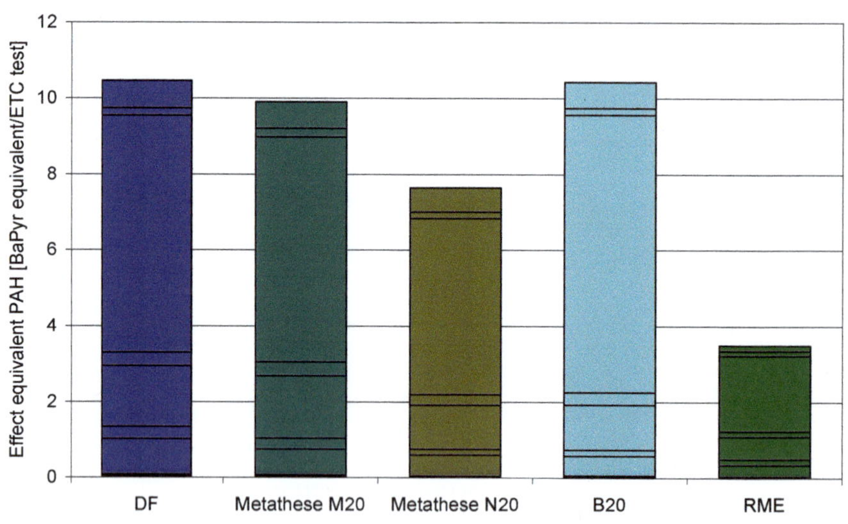

Fig. 4-36: Effect equivalent of the PAH emissions for particulate and condensate of the OM 904 LA in the ETC test

When the totals are based on the effect equivalent of the benzo[a]pyrene, Fig. 4-36 is produced. Here, in addition to the RME, the metathesis blend N20 also exhibits reduced effective potential.

However, if one considers the standard deviations, this behaviour cannot be reliably confirmed. On the other hand, the decrease is very clear for RME and is also present for almost all of the measured PAH.

In the fuel tests in the commercial vehicle engine, twelve substances from the class of carbonyls are identified for all fuels, where butanone and butyraldehyde are identified together for metrological reasons.

Fig. 4-37 shows the aldehyde emissions of the RME and the three blend fuels based on DF, which thereby corresponds with the 100% value in the illustration. Only very small and insignificant differences showed between the individual fuels, as before with the single-cylinder engine. Only the blend of the fuel MN produced by cross-metathesis with 1-hexene exhibited an increase in the case of some of the measured carbonyls. But even this increase was in the magnitude of the standard deviation.

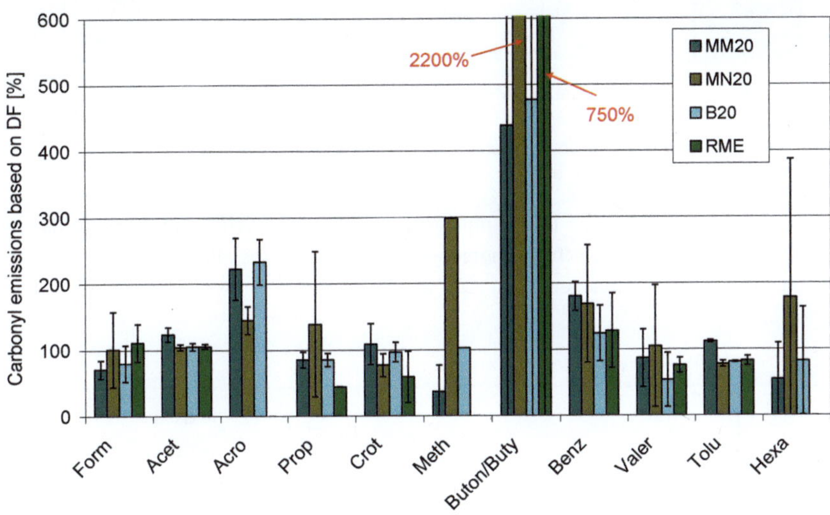

Fig. 4-37:   Comparison of carbonyl emissions of RME and three blend fuels of the OM 904 LA in the ETC test, based on DF

Triple determination was carried out for all fuels. However, some results are missing due to mistakes in the analytics. At these points, no bars are shown in the diagram. The absence of the standard deviation in the case of some measured values indicates that only one measurement could be used for the result.

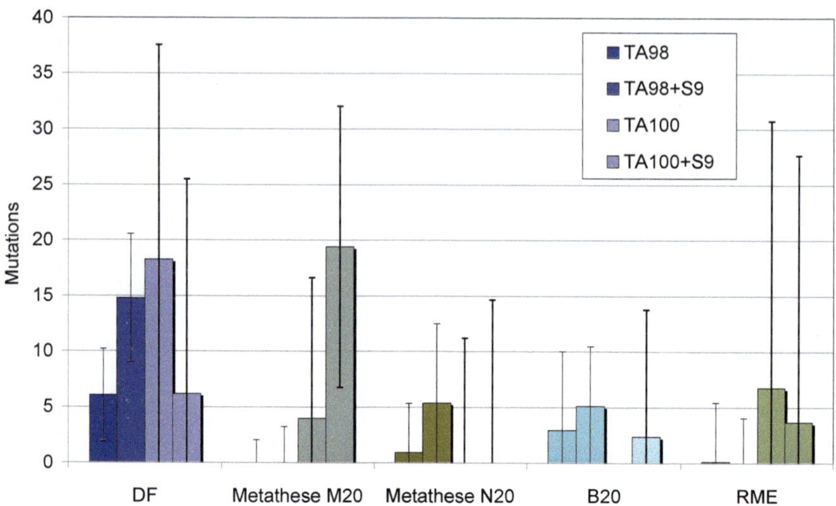

Fig. 4-38:   Comparison of the mutagenicity of particulates of DF and RME in comparison with the three blend fuels on the OM 904 LA in the ETC test

It is clear in Fig. 4-38 that almost no further mutagenic effects can be verified in the particulate of the exhaust gas of the commercial vehicle engine with SCR exhaust gas after-treatment. There are only occasional mutations but these predominantly do not differ from zero since the error indicators include zero.

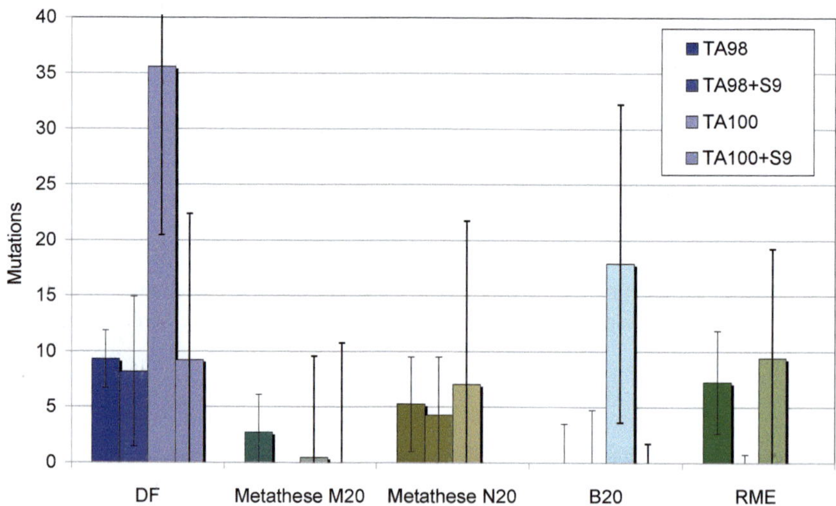

Fig. 4-39:   Comparison of mutagenicity of condensates of DF and RME in comparison with three blend fuels on the OM 904 LA in the ETC test

Here, no maximum of mutagenicity for the B20 blend can be verified with the utilised engine. As with the Farymann engine (cf. Fig. 4-21 and Fig. 4-22), this could be due to further developments in the area of the fuel filter and the resultant precipitation of oligomers. At best, diesel fuel tends to exhibit a small mutagenic potential. In addition, the results for the condensates presented in Fig. 4-39 show a small mutagenic potential for DF only.

The determined results clearly show that hardly any mutagenic potential could be verified in the exhaust gas for all five fuels. The addition of biofuels to the diesel fuel reduces the mutations per sample from up to 35 to values below the detection threshold in the magnitude of 15 to 20. With these values, the zero point of the standard deviation is included. Hence neither advantages nor disadvantages were exhibited by the fuel that was altered by metathesis.

4.4 Determination of the emissions and combustion behaviour

The results of the measurements on the AVL single-cylinder test engine from ivb can be found in below. The emissions and combustion behaviour is discussed by way of example using two selected operating points. The remaining measurements showed comparable trends.

In Fig. 4-40, the statutorily-regulated emissions and the indicated fuel consumption for operating points three and four (Table 3-6) are presented.

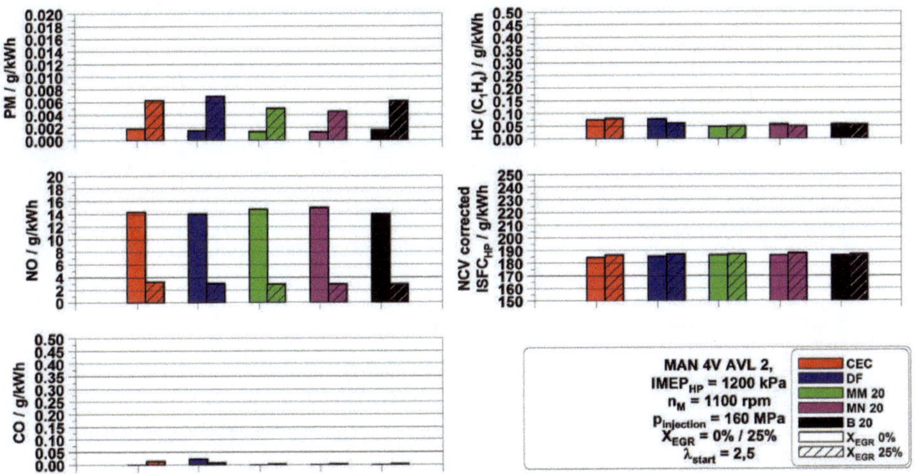

Fig. 4-40: Emissions and consumption for operating points three and four (Table 3-6) of the AVL single-cylinder test engine

Analogous to the other measurements, the two points only differ in that the first point is always operated as a starting point without exhaust gas recirculation and the second point with EGR represented the intended pre-series operating point. The only exception was operating point nine, which recreated a pre-series idle point and hence was generally applied without EGR.

It can be ascertained that the engine behaviour of all tested fuels is very similar and therefore the emissions behaviour corresponds to that of a conventional diesel combustion process.

Particle emissions at operating point three were very low for all fuels. With an EGR rate of 25% in operating point four, the particle emissions increased in accordance with expectations. In both points, the metathesis blends displayed somewhat lower particle emissions than the other fuels. This process could be observed for all measurement points.

The nitrogen oxide emissions showed almost no differences without EGR, however a small increase for the metathesis blends in combination with exhaust gas recirculation could be ascertained.

For almost all operating points, the carbon monoxide emissions are around the detection limit with 0 to 10 ppm since the measurement range was set out from 0 to 5000 ppm. Only operating points eight and nine showed slightly increased CO emissions.

The emissions behaviour of the non-combusted hydrocarbons exhibited slight advantages for the fuel blends with a high biogenic proportion. But this was also at a very low level.

The indicated fuel consumption exhibited the usual slight increase with increase in EGR. This is to be attributed to slowed combustion as a result of higher specific heat capacity and lower oxygen concentration in the cylinder charge. In accordance with expectations, all operating points exhibited slight consumption disadvantages for the fuel blends with a high biogenic proportion due to lower heating values of the biocomponents. However, compensation is exhibited due to corresponding heating value correction of the consumption figures.

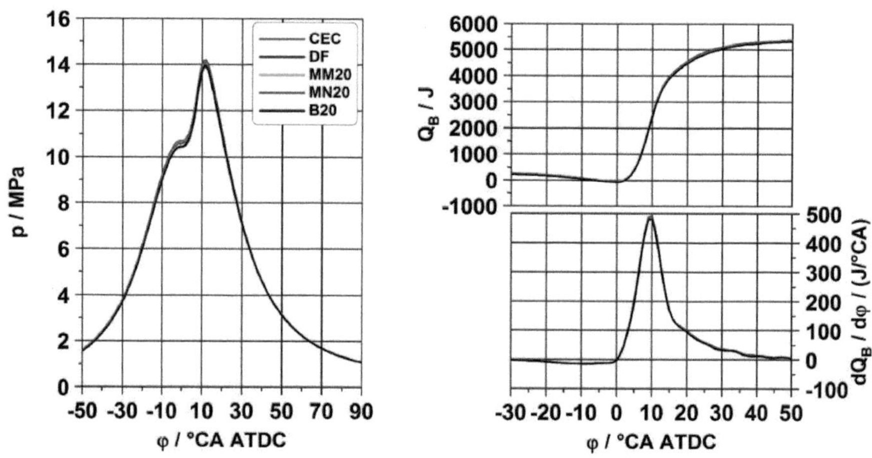

Fig. 4-41:   Thermodynamic evaluation of operating point three of the AVL single-cylinder test engine

The thermodynamic evaluations for operating points three and four, as just discussed, are presented in Fig. 4-41 and Fig. 4-42. Corresponding with the slight differences in emissions behaviour, the thermodynamic analyses also displayed barely-visible deviations in comparison with the different fuels.

Slight deviations in the pressure gradients can be explained by the different boost pressures for the same starting air ratio due to variations in the minimum air requirement. Differences in ignition behaviour and energy release could not be clearly discerned, indicating slight differences with regard to mixture formation as well as combustion and burning behaviour in the tested fuels.

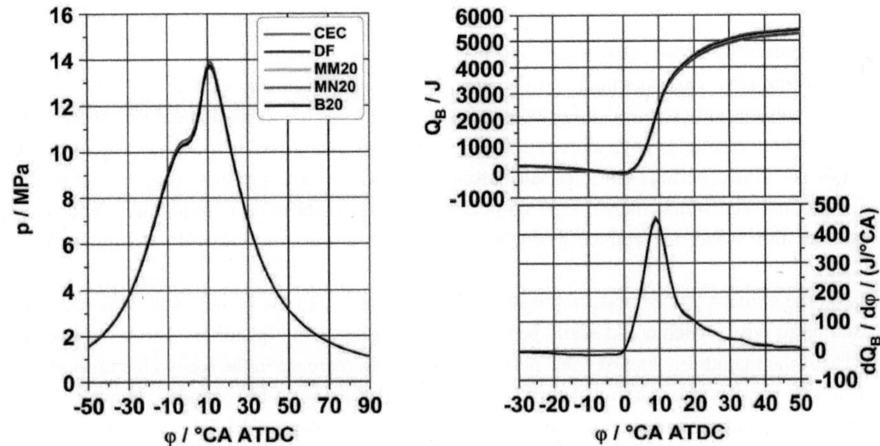

Fig. 4-42:   Thermodynamic evaluation of operating point four of the AVL single-cylinder test engine

5   Outlook

The results presented here clearly show that it is possible to change the boiling point of common commercial biofuels. This adaptation of the fuels allowed the disadvantage of oil dilution of biodiesel to be substantially reduced. In principle, it should be possible to implement this process on a large scale. Many examples have already shown that it is not only possible to technically implement metathesis but it also permits commercially-attractive processes and products (e.g. SHOP process, triolefin process and much more). One company that uses metathesis commercially and among other things also wants to produce biofuels is Elevance Renewable Sciences, Bolingbrook, Ill., USA.

Effects on the exhaust gas composition cannot be determined with the blends with 20% metathesis fuel used up to now, which could easily be the case for higher blend quantities. If one assumes that only 50% of the metathesis fuel remains in the engine oil, an increase in the renewable proportion in fuel would be conceivable without the oil change intervals having to be extended. However, in order to be able to make detailed assertions on this matter, further tests of the interaction between fuel and engine oil would be required.

The selected starting components of the metathesis reaction offer a further approach. In accordance with current renewable energy guidelines, biofuels must attain a reduction in greenhouse gases of at least 50% in order to be approved for the market as of 2017. However, at its current stage of development, biodiesel only attains a reduction of 38% (default value of guideline 2009/28/EC). Therefore it could be conceivable in future to apply the metathesis reaction not to ready-transesterified biodiesel but directly to rapeseed oil. In this way, it might be possible under certain circumstances to develop the possibility of making the rapeseed oil more accessible to the fuel market and, in addition, to still benefit from the advantages of an individually adjustable boiling curve for the fuel.

Bibliography

Ames, B.N., Lee, F.D., Durston, W.E. (1973). An improved bacterial test system for the detection and classification of mutagens and carcinogens. Proc Natl Acad Sci USA 70, 782 - 786

Ames, B.N., McCann, J., Yamasaki, E. (1975). Methods for detecting carcinogens and mutagens with the Salmonella/mammalian-microsome mutagenicity test. Mutat Res 31, 347 - 363

Bachler, C., Schober, S., Mittelbac M. (2010). Simulated Distillation for Biofuel Analysis. Energy Fuels, 24, 2086-2090

Baek, B.H., Aneja, V.P., Tong, Q. (2004). Chemical coupling between ammonia, acid gases, and fine particles. Environmental Pollution 129, 89 - 98

BAMS - Bundesministerium für Arbeit und Soziales (2006). Technische Regeln für Gefahrstoffe 900 - Arbeitsplatzgrenzwerte.

van Basshuysen, R., Schäfer, F. (2002). Handbuch Verbrennungsmotor - Grundlagen, Komponenten, Systeme, Perspektiven. Braunschweig/Wiesbaden, Friedr. Vieweg & Sohn Verlagsgesellschaft mbH, 2. verbesserte Auflage, 888 Seiten, ISBN 3-528-13933-1

Belisario, M.A., Buonocore, V., De Marinis, E., De Lorenzo, F. (1984). Biological availability of mutagenic compounds adsorbed onto diesel exhaust particulate. Mutat. Res. 135, 1–9

BioKraftQuG (2006). Gesetz zur Einführung einer Biokraftstoffquote durch Änderung des Bundes-Immissionsschutzgesetzes und zur Änderung energie- und stromsteuerrechtlicher Vorschriften. Bundesgesetz, BT-Drs 16/2709

Brooks, A.L., Wolff, R.K., Royer, R.E., Clark, C.R., Sanchez, A., McClellan, R.O. (1980). Biological availability of mutagenic chemicals associated with diesel exhaust particles; in Health Effects of Diesel Engine Emissions. EPA/600/9-80/57a, U.S. Environmental Protection Agency, Cincinnati, USA

BS ISO 16183:2002, (2002). Heavy-duty engines - Measurement of gaseous emission from raw exhaust gas and of particulate emission using partial flow dilution systems under transient test conditions

Bünger, J., Krahl, J., Prieger, K., Munack, A., Hallier, E. (1998). Mutagenic and cytotoxic effects of exhaust particulate matter of biodiesel compared to fossil diesel fuel. Mutat Res 415, 13-23

Bünger, J., Müller, M.M., Krahl, J., Baum, K., Weigel, A., Hallier, E., Schulz, T.G. (2000). Mutagenicity of diesel exhaust particles from two fossil and two plant oil fuels. Mutagenesis 15(5):391-39

Clark, C.R., Vigil, C.L. (1980). Influence of rat lung and liver homogenates on the mutagenicity of diesel exhaust particulate extracts. Toxicol. Appl. Pharmacol. 56, 100–115

Claxton, L.D., Barnes, H.M. (1981). The mutagenicity of diesel-exhaust particle extracts collected under smoke-chamber conditions using the Salmonella typhimurium test system. Mutat. Res. 88, 255–272

Claxton, L.D. (1983). Characterization of automotive emissions by bacterial mutagenesis bioassay: a review. Environ Mutagen 5, 609 - 631

Dekati Ltd. (2002). ELPI - User Manual. Version 3.13, Dekati Ltd., Tampere

Deutsche Forschungsgemeinschaft (1987). Gesundheitsschädliche Arbeitsstoffe, Toxikologisch-arbeitsmedizinische Begründungen von MAK-Werten: Dieselmotoremissionen. Wiley-VCH, Weinheim 1987

DIN EN ISO 175 (2010). Kunststoffe - Prüfverfahren zur Bestimmung des Verhaltens gegen flüssige Chemikalien, Beuth Verlag GmbH, Berlin

DIN EN ISO 2719 (2002). Bestimmung des Flammpunktes - Verfahren nach Pensky-Martens mit geschlossenem Tiegel

DIN EN ISO 527-2 (2010). Kunststoffe - Bestimmung der Zugeigenschaften - Teil 2: Prüfbedingungen für Form- und Extrusionsmassen, Beuth Verlag GmbH, Berlin

DIN EN 590:2009+A1:2010 (2010). Kraftstoffe für Kraftfahrzeuge - Dieselkraftstoff – Anforderungen und Prüfverfahren

EG 2005/55 (2005). Richtlinie 2005/55/EG des Europäischen Parlaments und des Rates vom 28. September 2005 zur Angleichung der Rechtsvorschriften der Mitgliedstaaten über Maßnahmen gegen die Emission gasförmiger Schadstoffe und luftverunreinigender Partikel aus Selbstzündungsmotoren zum Antrieb von Fahrzeugen und die Emission gasförmiger Schadstoffe aus mit Flüssiggas oder Erdgas betriebenen Fremdzündungsmotoren zum Antrieb von Fahrzeugen

EG 2009/28 (2009). RICHTLINIE 2009/28/EG DES EUROPÄISCHEN PARLAMENTS UND DES RATES vom 23. April 2009 zur Förderung der Nutzung von Energie aus erneuerbaren Quellen und zur Änderung und anschließenden Aufhebung der Richtlinien 2001/77/EG und 2003/30/EG

EG 582/2011 (2011). VERORDNUNG (EU) Nr. 582/2011 DER KOMMISSION vom 25. Mai 2011 zur Durchführung und Änderung der Verordnung (EG) Nr. 595/2009 des Europäischen Parlaments und des Rates hinsichtlich der Emissionen von schweren Nutzfahrzeugen (Euro VI) und zur Änderung der Anhänge I und III der Richtlinie 2007/46/EG des Europäischen Parlaments und des Rates

EG 595/2009 (2009). Verordnung (EG) Nr. 595/2009 des Europäischen Parlaments und des Rates vom 18. Juni 2009 über die Typgenehmigung von Kraftfahrzeugen und Motoren hinsichtlich der Emissionen von schweren Nutzfahrzeugen (Euro VI) und über den Zugang zu Fahrzeugreparatur- und -wartungsinformationen, zur Änderung der Verordnung (EG) Nr. 715/2007 und der Richtlinie 2007/46/EG sowie zur Aufhebung der Richtlinien 80/1269/EWG, 2005/55/EG und 2005/78/EG

Fernando S., Hall C., Jha S. (2006). NO_x reduction from biodiesel fuels. Energy and Fuels 20(1):376-382

Fuhrhop, J.-H., Li, G. (2003). Organic Synthesis, Concepts and Methods, VCH, Weinheim

Gekas, I., Gabrielsson, P., Johansen, K., Nyengaard, L., Lund, T. (2002). Urea-SCR Catalyst System Selection for Fuel and PM Optimized Engines and a Demonstration of a Novel Urea Injection System. SAE Technical Paper Series 2002-01-0289

GESTIS-Stoffdatenbank (2011). Gefahrstoffinformationssystem der gewerblichen Berufsgenossenschaften. http://www.hvbg.de (zitiert am 12.04.2011)

Hackbarth, E.-M., Merhof, W. (1998). Verbrennungsmotoren – Prozesse, Betriebsverhalten, Abgas. Braunschweig, Wiesbaden, Friedrich Vieweg & Sohn Verlagsgesellschaft mbH, 137 Seiten, ISBN 3-528-07431-0

Huisingh, J., Bradow, R., Jungers, R., Claxton, L., Zweidinger, R., Tejada, S., Bumgarner, J., Duffield, F., Waters, M. (1978). Application of bioassay to the characterization of diesel particle emissions; in: Application of short-term bioassay in the fractionation and analysis of complex environmental mixtures; hrsg. v. Waters M.D., Nesnow S., Huisingh J.L., Sandhu S.S., Claxton L.D.; Plenum Press, New York, 382 - 418

Krahl, J., Baum, K., Hackbarth, U., Jeberien, H.-E., Munack, A., Schütt, C., Schröder, O., Walter, N., Bünger, J., Müller, M.M., Weigel, A. (2001). Gaseous compounds, ozone precursors, particle number and particle size distributions, and mutagenic effects due to biodiesel. Transactions of the ASAE 44(2):179-191

Krahl, J., Munack, A., Schröder, O., Stein, H., Bünger, J. (2003). Influence of biodiesel and different designed diesel fuels on the exhaust gas emissions and health effects. SAE Technical Paper Series 2003-01-3199

Krahl, J., Munack, A., Ruschel, Y., Schröder, O., Bünger, J. (2008). Exhaust Gas Emissions and Mutagenic Effects of Diesel Fuel, Biodiesel and Biodiesel Blends. SAE Technical Paper Series 2008-01-2508

Krahl, J., Petchatnikov, M., Schmidt, L., Munack, A., Bünger, J. (2009). Spektroskopische Untersuchungen zur Ergründung der Wechselwirkungen zwischen Biodiesel und Dieselkraftstoff bei Blends. Abschlussbericht zum Forschungsvorhaben 7-TA-VDB, Johann Heinrich von Thünen-Institut, Braunschweig

Krahl, J., Schmidt, L., und Munack, A. (2011). Untersuchungen zu biodieselbasierten Mischkraftstoffen mit geringer Neigung zur Präzipitatbildung sowie zur Verwendung von Ethanol als Komponente in Dieselkraftstoff-Biodieselmischungen. Abschlussbericht zum Forschungsvorhaben 550_2009_2, Johann Heinrich von Thünen-Institut, Braunschweig

Krewski, D., Leroux, B.G., Creason, J., Claxton, L. (1992). Sources of variation in the mutagenic potency of complex chemical mixtures based on the Salmonella/microsome assay. Mutat Res. 1992 276:33-59.

Lapuerta, M., Armas, O., Rodrıguez-Fernandez, J. (2008). Effect of biodiesel fuels on diesel engine emissions. Progress in Energy and Combustion Science, 34, 198–223

Lewtas, J. (1983). Evaluation of the mutagenicity and carcinogenicity of motor vehicle emissions in short-term bioassays. Environ. Health Perspect 47, 141-152

Matsushima, T., Sawamura, M., Hara, K., Sugimura, T. (1976). A safe substitute for polychlorinated biphenyls as an inducer of metabolic activation system; in: In vitro metabolic activation in mutagenesis testing; hrsg. v. Serres FJ, Fouts JR, Bend JR, Philpot RM; Elsevier/North-Holland, Amsterdam, 85-88

Matsushita, H., Goto, S., Endo, O., Lee, J., Kawai, A. (1986). Mutagenicity of diesel exhaust and related chemicals, in: Carcinogenic and Mutagenic Effects of Diesel Engine Exhaust. Elsevier Science Publishing, New York, 103-118

Miao, X., Malacea, R., Fischmeister, C., Bruneau, C., Dixneuf, P.H. (2011). Ruthenium–alkylidene catalysed cross-metathesis of fatty acid derivatives with acrylonitrile and methyl acrylate: a key step toward long-chain bifunctional and amino acid compounds, Green Chem., 13, 2911-2919

Mollenhauer, K. (2002). Handbuch Dieselmotoren. Berlin, Springer-Verlag, 2. korrigierte und neu bearbeitete Auflage, 1069 Seiten, ISBN 3-540-41239-5

Mortelmans, K., Zeiger, E. (2000). The Ames Salmonella/microsome mutagenicity assay. Mutat Res. 2000 455:29-60

Ngo, H., Jones, K., Foglia, T. (2006). Metathesis of Unsaturated Fatty Acids: Synthesis of Long-Chain Unsaturated-α-,ω-Dicarboxylic Acids, J. Am. Oil Chem. Soc., **83**, 629-634

Nold, A., Bochmann, F. (1999). Epidemiologische Ergebnisse zu Dieselmotoremissionen und Lungenkrebs: Eine Synopse. Gefahrstoffe - Reinhaltung der Luft, 59, 289 – 298

Ohe, T. (1984). Mutagenicity of photochemical reaction products of polycyclic aromatic hydrocarbons with nitrite. Sci. Total. Environ. 39, 161–175

Pederson, T.C., Siak, J.S. (1981). The role of nitroaromatic compounds in the direct-acting mutagenicity of diesel particle extracts. J Appl Toxicol 1, 54 - 60

Rannug, U., Sundvall, A., Westerholm, R., Alsberg, T., Stenberg, U. (1983). Some aspects of mutagenicity testing of the particulate phase and the gas phase of diluted and undiluted automobile exhaust. Environ. Sci. Res. 27, 3–16

Rosenkranz, H.S., Mermelstein, R. (1983). Mutagenicity and genotoxicity of nitroarenes. All nitro-containing chemicals were not created equal. Mutat. Res. 114, 217–267

Savela, K., Pohjola, S., Lappi, M., Rantanen, L. (2003). Polycyclic aromatic hydrocarbons derived from vehicle air pollutants in work environment and exposure to bronchus epithelial cell line (BEAS-2B). In: 7th ETH Conference on Combustion Generated Particles from 18th to 20th August 2003 in Zürich, Tagungsband, CD-ROM

Scheepers, P.T.J., Bos, R.P. (1992). Combustion of diesel fuel from a toxicological perspective, II. Toxicity. Int. Arch. Occup. Environ. Health, 64, 163 - 177

Schröder, O., Baum, K., Hackbarth, U., Krahl, J., Prieger, K., Schütt, C. (1998). Einfluss von Gemischen von Dieselkraftstoff und Biodiesel auf das Abgasverhalten. Fachtagung Biodiesel - Optimierungspotentiale und Umwelteffekte. Braunschweig, p. 147-154

Shirnamé-Moré, L. (1995). Genotoxicity of diesel emissions; Part I: Mutagenicity and other genetic effects; in: Diesel exhaust: A critical analysis of emissions, exposure, and health effects; hrsg. v. Health Effects Institute, Cambridge, USA, 223 - 242

Siak, J.S., Chan, J.L., Lee, P.S. (1981). Diesel particulate extracts in bacterial test systems. Environ. Int. 5, 243-248

Skaanderup, P.R., Jensen, T. (2008). Synthesis of the Macrocyclic Core of (-)-Pladienolide B, Org. Lett., 10 (13), 2821-2824

Stein, H. (2008). Dieselmotoremissionen aus der Verbrennung von Biodiesel und verschiedenen fossilen Dieselkraftstoffen unter besonderer Berücksichtigung der Partikelemissionen. Dissertation, Braunschweig

Stump, F., Bradow, R., Ray, W., Dropkin, D., Zwedinger, R., Sigsby, J., Snow, R. (1982). Trapping gaseous hydrocarbons for mutagenic testing. Paper No. 820776; Society of Automotive Engineers, Warrendale, USA

Tschöke, H., Braungarten, G., Patze, U. (2008). Ölverdünnung bei Betrieb eines Pkw-Dieselmotors mit Mischkraftstoff B10. Abschlussbericht, OTTO-VON-GUERICKEUNIVERSITÄT, Institut für Mobile Systeme - Lehrstuhl Kolbenmaschinen, Magdeburg

UFOP (2010). Biodiesel 2009/2010 Sachstandsbericht und Perspektive – Auszug aus dem UFOP-Jahresbericht, UNION ZUR FÖRDERUNG VON OEL- UND PROTEINPFLANZEN E. V., Berlin

Umweltbundesamt (2003). Bericht der Bundesrepublik Deutschland nach Art. 6 der Richtlinie 2001/81/EG (NEC-Richtlinie) über die Emissionen von SO_2, NO_x, NH_3 und NMVOC sowie die Maßnahmen zur Einhaltung der NECs. www.umweltbundesamt.de/luft/downloads/luftreinhalte-strategien /NECAnhang1.pdf [zitiert am 10.04.2011]

Umweltbundesamt (2010). Nationale TrendTablen für die deutsche Berichterstattung atmosphärischer Emissionen. 1990 - 2008 (Fassung zur EU-Submission 15.01.2010)

VDA Verband der Automobilindustrie e.V. (2004). Auto Jahresbericht 2004. www.vda.de (zitiert am 10.04.2011)

VDA Verband der Automobilindustrie e.V. (2010). Jahresbericht 2010. www.vda.de (zitiert am 10.04.2011)

VDI (Verein Deutscher Ingenieure, 1989). Messen von Emissionen/Messen von polycyclischen aromatischen Kohlenwasserstoffen (PAH)/Messen von PAH in Abgasen von PKW-Otto- und -Dieselmotoren/Gaschromatographische Bestimmung. VDI 3872, Blatt 1. Beuth-Verlag, Berlin.

V&F Analyse- und Messtechnik GmbH (2011). Process Mass Spectrometer Model AIRSENSE.net. www.vandf.com (zitiert am 25.01.2011)

V&F Analyse- und Messtechnik GmbH (2011). Technische Beschreibung AIRSENS.net, Version 08/07, Absam

Wang, Y.Y., Rappaport, S.M., Sawyer, R.F., Talcott, R.E., Wei, E.T. (1978). Direct-acting mutagens in automobile exhaust. Cancer Lett 5, 39 - 47

Warnatz, J., Maas, U., Dibble, R.W. (2001). Verbrennung - physikalisch-chemische Grundlagen, Modellierung und Simulation, Experimente, Schadstoffentstehung. Berlin, Springer-Verlag, 3. aktualisierte und erweiterte Auflage, 326 Seiten, ISBN 3-540-42128-9

Welschof, G. (1981). Der Ackerschlepper - Mittelpunkt der Landtechnik. VDI-Berichte 407. S. 11-17

6 Appendix

6.1 Fuel analyses

The analysis of the RME used in the project was carried out upon procurement of the fuel in May 2009. Hence for some values, for example oxidation stability, it is not applicable to the entire period of use of the fuel. The oxidation stability was tested again four months after completion of the last measurements. This measured value, which is also listed in the table, clearly lies below the standard specification.

However, at no point in the project was the formation of turbidity or sediment observed caused by oligomer formation due to fuel aging when producing the fuel mixtures.

Property	Method	Unit	Threshold values		RME
Analysis of 05/2009					
			Min.	Max.	
Ester content	DIN EN 14103	Wt. %	96.5		98.0
Density (15 °C)	DIN EN ISO 12185	kg/m^3	875	900	883.4
Kin. viscosity (40 °C)	DIN EN ISO 3104	mm²/s	3.5	5.0	4.430
Flash point	DIN EN ISO 3679	°C	120		170
CFPP	DIN EN 116	°C		0/-10/-20	-16
Sulphur content	DIN EN ISO 20884	mg/kg		10.0	< 10
Coke residue	DIN EN ISO 10370	Wt. %		0.3	< 0.30
Cetane number	IP 498		51		> 51
Ash content	ISO 3987	Wt. %		0.02	< 0.01
Water content	DIN EN ISO 12937	mg/kg		500	203
Total contamination	DIN EN 12662	mg/kg		24	1
Corrosive effect on copper	DIN EN ISO 2160	Degree of corrosion		1	1
Oxidation stability	DIN EN 14112	h	6		> 8
Acid number	DIN EN 14104	mg KOH/g		0.5	0.11
Iodine number	DIN EN 14111	g iodine/100 g		120	114
Linolenic acid methyl content.	DIN EN 14103	Wt. %		12	10.4
Methanol content	DIN EN 14110	Wt. %		0.20	0.02
Free glycerin	DIN EN 14105	Wt. %		0.02	< 0.005
Monoglycerides	DIN EN 14105	Wt. %		0.80	0.59
Diglycerides	DIN EN 14105	Wt. %		0.20	0.14
Triglycerides	DIN EN 14105	Wt. %		0.20	0.07
Total glycerine content	DIN EN 14105	Wt. %		0.25	0.18
Phosphor content	DIN EN 14107	mg/kg		10	< 1
Alkali content	E DIN EN 14538	mg/kg		5	< 1
Alkaline earth metal content	E DIN EN 14538	mg/kg		5	< 1
Fatty acid methyl ester content >/= 4 double bonds		%		1	< 1
Oxidation stability 05/2012	DIN EN 15751	h	6		< 0.5

Table 6-1:   Analysis of rapeseed oil methyl ester

Composition of the RME					
12:0 lauric acid					< 0.1
14:0 myristic					< 0.1
16:0 palmitic					4.7
16:1 palmitoleic					0.3
18:0 stearic					1.6
18:1 oleic					61.2
18:2 linoleic					19.2
18:3 linolenic					10.1
20:0 arachidic					0.5
20:1 eicosenoic					1.3
22:0 behenic					0.3
22:1 erucic					0.3
24:0 lignoceric					0.2
24:1 nervonic					0.2
Total					99.9

Table 6-2:   Fatty acid spectrum of the utilised RME

Property	Method	Unit	Threshold values		DK
			Min.	Max.	
Cetane number	DIN EN ISO 5165	-	49	-	53.4
Density (15 °C)	DIN EN ISO 12185	kg/m^3	820	845	834.3
Polycycl. aromatic HC	DIN EN 12916/IP 391	% (m/m)	-	11	4.6
Sulphur content	DIN EN ISO 20884	mg/kg	-	50	3
Flash point	DIN EN ISO 22719	°C	> 55	-	92
Coke residue	DIN EN ISO 10370	% (m/m)	-	0.30	<0.01
Ash content	DIN EN ISO 6245	% (m/m)	-	0.01	<0.001
Water content	DIN EN ISO 12937	mg/kg	-	200	23
Corrosive effect on copper	DIN EN ISO 2160	Corr. degree	-	1	1A
Oxidation stability	DIN EN ISO 12205	g/m^3	-	25	< 1
HFRR (at 60°C)	DIN EN ISO 12156-1	μm	-	460	235
Kin. viscosity (40 °C)	DIN EN ISO 3104	mm²/s	2	4.5	3.126
CFPP	DIN EN 116	°C	-	0/-10/-20	-17
Distillation course					
95% point	DIN EN ISO 3405	°C	-	360	347.7
FAME content	DIN EN ISO 14078	% (V/V)	-	5	passed
Hydrogen	(ASTM D 3343)	%			13.72
Carbon	(ASTM D-3343)	%			86.28
Heating value	(ASTM D 3338)	MJ/kg			43.226
Cloud point	EN 23015	°C			-15
Neutralisation number	(ASTM D 974)	mg KOH/g			< 0.02
Monoaromatics	(IP 391)	% (m/m)			15.4
Diaromatics	(IP 391)	% (m/m)			4.6
Tri+further	(IP 391)	% (m/m)			< 0.1
Polyaromatics*,**		% (m/m)			4.6
Total aromatics	(IP 391)	% (m/m)			20

Table 6-3:   Analysis of the reference diesel fuel from vTI

Property	Method DIN EN	Unit	Threshold values		Test result
			Min.	Max.	
Cetane number	ISO 15195	-	51.0	-	55.7
Cetane index	ISO 4264	-	46.0	-	55.1
Density (15 °C)	ISO 12185	kg/m^3	820	845	838.8
Polycycl. aromatic HC	ISO 12916	% (m/m)	-	8.0	3.5
Sulphur content	ISO 20884	mg/kg	-	10	1.0
Flash point	ISO 2719	°C	> 55	-	79.0
Coke residue	ISO 10370	% (m/m)	-	0.30	0.25
Ash content	ISO 6245	% (m/m)	-	0.01	< 0.005
Water content	ISO 12937	mg/kg	-	200	233
Total contamination	12662	mg/kg	-	24	8
Corrosive effect on copper	ISO 2160	Corr. degree	Class 1		1
Oxidation stability	15751	h	20	-	< 0.5
HFRR (at 60 °C)	ISO 12156-1	μm	-	460	266
Kin. viscosity (40 °C)	ISO 3104	mm²/s	2.0	4.5	3.051
CFPP	116	°C	-	0/-10/-20	-17
Distillation course					
% (V/V) 250 °C	ISO 3405	% (V/V)	-	<65	20.8
% (V/V) 350 °C	ISO 3405	% (V/V)	85	-	94.3
95% point	ISO 3405	°C	-	360	353.7
FAME content	ISO 14078	% (V/V)	-	7.0	18.1

Table 6-4: Analysis of metathesis fuel blend MM20 (ASG company)

Test parameter	Method	Unit	Threshold values DIN EN 14214:10-04		Test result
			min.	max.	
Ester content	DIN EN 14103	% (m/m)	96.5	-	45.5
Density at 15 ^{o}C	DIN EN ISO 12185	kg/m^3	860	900	881.0
Kin. viscosity at 40 ^{o}C	DIN EN ISO 3104	mm^2/s	3.5	5.0	5.268
Flash point	DIN EN ISO 3679	^{o}C	101	-	118.0
CFPP	DIN EN 116	^{o}C	-	**	-8
Sulphur content	DIN EN ISO 20884	mg/kg	-	10.0	1.9
Coke residue (10% D.)	DIN EN 10370	% (m/m)	-	0.3	0.46
Cetane number	DIN EN 15195	---	51.0	-	59.1
Ash content (sulphate ash)	ISO 3987	% (m/m)	-	0.02	< 0.01
Water content K.-F.	DIN EN ISO 12937	mg/kg	-	500	1079
Total contamination	DIN EN 12662	mg/kg	-	24	< 1
Copper corrosion	DIN EN ISO 2160	Corr. degree	1		1
Oxidation stability	DIN EN 14112	h	6.0	-	< 0.5
Acid number	DIN EN 14104	mg KOH/g	-	0.5	0.493
Iodine number	DIN EN 14111	g iodine/100 g	-	120	59
Linolenic acid ME content	DIN EN 14103	% (m/m)	-	12.0	1.7
Methanol content	DIN EN 14110	% (m/m)	-	0.20	< 0.01
Content of free glycerine		% (m/m)	-	0.020	< 0.01
Monoglyceride content		% (m/m)	-	0.80	0.32
Diglyceride content	DIN EN 14105	% (m/m)	-	0.20	0.06
Triglyceride content		% (m/m)	-	0.20	0.05
Total glycerine content		% (m/m)	-	0.25	0.10
Phosphor content	DIN EN 14107	mg/kg	-	4.0	< 0.05
Alkali content (Na + K)	DIN EN 14108/109	mg/kg	-	5.0	< 0.05
Alkaline earth metal content (Ca + Mg)	DIN EN 14538	mg/kg	-	5.0	< 0.05
Ruthenium content	ICP-OES	mg/kg	-	-	6

Table 6-5: Analysis of a self-metathesis sample of biodiesel (0.05 mol% Umicore M5$_1$) in accordance with DIN 14214 (ASG company)